Praise for *Digital Body Language*:

"Non-verbal cues are vital to understanding each other. Now that so much communication happens online—and with the massive shift to distance learning and remote workplaces—we need Erica Dhawan's book more than ever. In *Digital Body Language*, Erica shares tips and strategies for communicating effectively on chats, emails, and video calls, so everyone can successfully share and connect in the digital age."

> Sheryl Sandberg, COO of Facebook and
> founder of Lean In and Option B

"*Digital Body Language* is an indispensable guide to a business world turned upside down by video calls, group texts, and remote work. With Dhawan's expert guidance, you'll learn how to read and send the subtle cues that signal trust, competence, and authenticity. You'll discover practical tips for using everything from exclamation points to emojis. Most of all, you'll understand that effective communication and collaboration begin with valuing others."

> Daniel H. Pink, author of *When*, *Drive*, and
> *To Sell Is Human*

"A profound look at how to foster inclusion and better leadership in our digital world. *Digital Body Language* makes a strong case for the importance of equality in all forums."

> Billie Jean King, Founder, Billie Jean King Leadership Initiative

"This book is a breakthrough that will be read for years to come. People are already judging you by how you show up in the digital world, but now Erica Dhawan helps you figure out how to earn the benefit of the doubt."

> Seth Godin, Author, *The Practice*

"Little have we yet recognized how the new virtual world of our communications can potentially undermine our good intentions, and this book has intelligently and elegantly described its challenges and offered the solutions. If you're e-mailing, zooming, IM-ing, etc., in your important relationships and for getting things done, professionally and personally, this is a must-read."

David Allen, international bestselling author of
Getting Things Done

"The first book to codify a new set of rules for succeeding in our modern work world. *Digital Body Language* is an essential workplace requirement."

Marshall Goldsmith, author of
What Got You Here Won't Get You There

DiGITAL BODY LANGUAGE

HOW TO BUILD TRUST & CONNECTION NO MATTER THE DISTANCE

ERiCA DHAWAN

HarperCollins*Publishers*

The names and identifying characteristics of some
persons described in this book have been changed.

HarperCollins*Publishers*
1 London Bridge Street
London SE1 9GF

www.harpercollins.co.uk

HarperCollins*Publishers*
1st Floor, Watermarque Building, Ringsend Road
Dublin 4, Ireland

First published in the US by St. Martin's Press, an imprint of
St. Martin's Publishing Group
This UK edition published by HarperCollins*Publishers* 2021

1 3 5 7 9 10 8 6 4 2

Design by Meryl Sussman Levavi

A catalogue record of this book is
available from the British Library

ISBN 978-0-00-847652-6

Printed and bound in Great Britain by
CPI Group (UK) Ltd, Croydon

MIX
Paper from
responsible sources
FSC™ C007454

To Kimaya and Rohan,
for inspiring me to stay forever curious.

To Rahul, for believing in me always.

ALSO BY ERICA DHAWAN

Get Big Things Done:
The Power of Connectional Intelligence
(with Saj-nicole Joni)

Contents

Introduction

After co-authoring my first book, *Get Big Things Done: The Power of Connectional Intelligence*, I traveled the world, speaking and consulting with companies and leaders on the challenges of twenty-first-century collaboration. My mission was to help CEOs and organizations solve the practical challenges of managing across global, multigenerational, matrixed, and virtual teams.

Everywhere I went, the same questions kept coming up: How do I keep my teams feeling connected to each other and to people on other teams? How do I help people of different ages and working styles who rarely meet in person communicate effectively? Why does it seem infinitely harder to foster trust, engagement, and the confidence to take risks? And finally, why do my own communications so often seem to miss the mark, producing unintended and anxiety-filled consequences?

The more I worked with my clients to solve these problems, the more obvious it became that they were caused by the very digital tools that had set us free in so many ways. Our failure to grapple with the communication-altering side effects of our shiny new digital tools—email, text messaging, Power-Point, Zoom—created widespread misunderstanding and conflict, which in turn manifested as across-the-board anxiety, fear, distrust, and paranoia.

The good news is that our communication problems are eminently solvable with some attention to a skill I call digital body language. I have taught many leaders how to model digital body language for their teams and how to introduce it to their cultures, with remarkable results. I have trained managers, HR teams, and coaches how to embed digital body language skills into their leadership programs. And I have advised everyone from doctors using telehealth to professors using online learning platforms to lawyers, consultants, and board directors using virtual meetings how to master this skill. One leader told me that simple changes in digital body language not only transformed the communication in her entire organization but also enhanced the customer experience that she was able to provide from afar. Another executive told me it changed how he connected with his wife and children while traveling for business.

Now I'll show you how digital body language can help *you*.

• • • • • •

When people ask me how I started doing what I do, I tell them it's a story that's lasted my whole life.

As a first-generation American girl born to Indian parents, I came to English in an indirect way. I grew up in a middle-class neighborhood outside of Pittsburgh. At home, my parents, both physicians who had immigrated to the United States in their twenties, spoke Punjabi—a language close to Hindi—and only rarely English. My mom and dad both made it a high priority that my two siblings and I honored traditional Indian values and customs. Silence was a sign of respect to elders, and listening a prized trait. Learning English, doing well academically, and almost everything else came in second.

Growing up in a white, conservative, suburban part of the country, I spent a lot of my childhood trying to fit in. There weren't many girls who looked like me, or who were the children of immigrants, or who sat down for dinner every night at 9:00 p.m. (Indian families tend to eat late.) At the same time, I felt almost no allegiance to India. Whenever I visited, my family there referred to me as "the American-born cousin." Who in India was named *Erica*?

Caught between two cultures, I went inside myself.

Often you would hardly know if I was in the room. In school I was shy, quiet, more an observer than a participant. Raising my hand or calling any attention to myself was unimaginable to me. I did well in school and on tests, but the comments in every report card I received from kindergarten to twelfth grade said the same thing: *I wish Erica spoke up more.*

As a girl pivoting between the thickly accented English of my parents and my own bad Hindi and wanting to feel like I belonged *somewhere*, I developed a few tricks, one of which was the ability to decipher other people's body language. Body language offered the key to understanding the foreign worlds around me. I became obsessed with decoding my classmates' signals and cues, no matter how subtle. Tone, pacing, pauses, gestures. The popular girls walked around with their heads high, shoulders pulled back, almost literally looking down on the rest of us. The older kids showed their disinterest by slouching during school assemblies, their eyes turned to the ground or each other—never to the adult speaking. At home, I holed up in my room watching Bollywood movies on my family's old VCR, focusing on the actors' faces and hands instead of on the storyline (Hindi was still alien to me), rewinding over and over again, trying to understand what was being said by observing the actors' nonverbal cues.

My preoccupation with translating nonverbal cues soon became a source of power as I learned to mimic the body language of my more confident peers and decode what my Hindi-speaking family members were saying to me with their furrowed brows.

After September 11, 2001, virtually everyone who looked like me in America was suddenly treated with instant suspicion in public spaces. One afternoon around that time, my father was waiting to pick me up at the local YMCA after tennis practice. Someone behind the front desk panicked—my dad "looked suspicious," I guess—and called the police. For the next 45 minutes, my father fielded questions from the officers, politely explaining that he worked as a cardiologist at a nearby hospital. I looked on as he sat behind a table, speaking patiently with the officers—his eye contact direct, his palms wide open, signaling his respect to the officers and his understanding of why this was happening. I could also tell by his flushed cheeks that he was embarrassed. A few months later, my dad donated a significant percentage of his income that year to the 9/11 fund.

I remember being angry at the police, but also at my father. How could he respond with kindness to what I saw as racial profiling and ignorance? Patiently, my father asked my siblings and me, wouldn't it be better to consider what other people might be thinking and feeling instead of responding with indignation or rage? To put ourselves in *their* shoes? That was an inflection point for me, a day I began thinking harder about how humans convey empathy via body language and what it can accomplish.

My interest in nonverbal communication continued in college, where I read every book on the subject I could find; later, I called on my growing expertise professionally when I began teaching public speaking. Being able to understand and classify cues and signals, along with the poise and confidence this skill gave me, helped me win internships and, eventually, ridiculously competitive job opportunities. All of this despite my father's insistence that Indian-Americans couldn't succeed in business and that I should instead focus on an occupation like medicine, or engineering, where Indians at least had a *tradition* of success. But I persevered—and it seemed to pay off.

My preoccupation with body language gave me the confidence to teach courses in leadership as a graduate student and, later, as a teaching fellow at both Harvard and MIT. It motivated me to start my own business before I was 30, which I was able to scale from a *what-if?* idea into a global company despite having no clue what I was doing, or any media experience, investors, or connections. Before I knew it, I was addressing global leaders at the World Economic Forum, getting interviewed by *Good Morning America* host Robin Roberts, becoming a "sought-after" keynote speaker by CEOs and top executives, and teaching twenty-first-century collaboration skills to thousands of leaders across a range of industries, companies, and countries.

If it sounds like I'm bragging, *please*! I hope it's clear by now that what I consider my "expertise" comes from shyer, humbler beginnings. There's a lot to be said for being withdrawn, refusing to raise your hand in class, and watching Bollywood movies alone in the dark after school. The point is, my whole life I've believed, as many do, that the essence of empathy and trust isn't about *what* we say but *how* we say it, and how often we check ourselves to make sure that our words and their meanings are as deliberate and clear as possible. Studying other people's body language as well as my own has taught me a lot, although the practical application of it has come through quite a bit of trial and error—in my case mostly error.

Hadn't my own experience taught me that my poor posture and dead-fish handshake made a negative impression on prospective employers? Hadn't one teacher told me that my habit of nervously twirling my hair signaled my insecurity? Hadn't I found out that a professor's tightly set lips or tensed nostril signaled whether I had aced or blown a test or paper? As a speaker, didn't I know that the difference between success and failure was about intuiting what an audience wanted and adjusting my message accordingly?

Once, early in my career, I delivered a keynote presentation in front of a big audience. It was the weekend, and the fourth day of an all-firm lawyer retreat. Not surprisingly, the audience was tired, peevish, disengaged, and tuned out. Some people looked overtly hostile. Others were slumped in their chairs, heads hanging to the side, eyes glancing up at the clock. The last things anyone wanted to hear about that day were the advantages of collaboration. Their body language almost begged me: *Not another framework, please.*

So I pivoted. Removing my heels, I took a seat on the edge of the stage, and scrapped my usual introduction. "Talk about the emotions you're feeling right now," I said. "Fatigue, tension, boredom, anticipation, rage, you name it . . ." Well, the mood in the room changed, just like that. I was no longer talking *at* the audience; I was talking *with* them. Everyone started to loosen up, relax, smile, laugh. A speech that could have been a disaster turned into an interactive hour of genuine connection and animated discussion.

Over the next few years I began to make it a point, as I do today, to suss out the body language of audiences. Blank expressions mean that I'm going too fast and need to slow down. Crossed arms signal defensiveness or resentment. As for me, I know that gesturing or adjusting my hair too much signals my own lack of confidence.

This brings me to a few years ago, when I started hearing one story after the next all centering on the same theme: *miscommunication in the workplace.*

As I said, I'd been giving keynote speeches and consulting with clients around the world, teaching people how to collaborate better at work. The most common questions I received were: How can we innovate faster and further by capitalizing on the expertise of digitally fluent employees while still leveraging an experienced workforce that is set in its ways? And how can I get these two groups to truly collaborate with each other? More and more clients and audience members of all ages were expressing high levels of fear, anxiety, and paranoia about communication in their workplace. Leaders were doing what they'd always done—for example, sharing messages of support and trust with their colleagues and teams—but more and more of those messages were being misunderstood, misinterpreted, or missed altogether. These leaders weren't dumb or lacking in social skills, and many were conversant in cutting-edge methods of building strong cultures.

As I dug more deeply into the responses I was getting, the biggest complaints seemed to revolve around how communications were being *translated* inside those same workplaces. That is, how a message that was meant to be friendly and to the point could be read by the recipient as angry or resentful, causing less engagement and innovation and even the loss of top performers.

This issue was illustrated by a meeting I had with a client, a senior leader at Johnson & Johnson whom I'll call Kelsey, who had gotten some tough feedback from her team on morale issues. In Kelsey's performance review, her boss commented that her "empathy was weak." When Kelsey and I first met and began talking, I kept my eye out for the standard, universal markers of subpar empathy: an inability to understand the needs of others, a lack of proficiency in reading and using body language, poor listening skills, a failure to ask deep questions. I was confused. Kelsey seemed to have *fantastic* empathy skills. She made me feel at ease, her body language signaled respect and understanding, and she listened deeply and carefully. What was going on?

The answer had less to do with Kelsey and more to do with today's tech-reliant workplace. Instead of *lacking* empathy, Kelsey, like nearly everyone I counseled, didn't know what empathy *meant* anymore in a world where digital communication had made once-clear signals, cues, and norms almost unintelligible. A tone of voice? Approachable body language? Those things didn't cut it anymore. The digital world required a *new* kind of body language. The problem was that no one could agree on what even *made up* that kind of body language.

For example, Kelsey believed she was doing everyone a favor by keeping her emails brief. But her team found them cold and ambiguous. Kelsey sent calendar invites at the last minute with no explanation, which made her teammates feel disrespected, as though Kelsey's schedule mattered more than theirs. During strategy presentations, Kelsey would glance down repeatedly at her phone, making others feel like she had checked out.

Kelsey's *digital* body language, then, was abysmal. It canceled out the very real clarity that comes when workplace colleagues (okay, humans in general) feel connected to one another via physical body language.

I realized that our understanding of body language needed to be redefined for the contemporary workplace. Today we're all "immigrants" learning a new culture and language, except this time it's in the digital space. Being a good leader today means not only being aware of other people's signals and cues but also mastering this new digital body language that didn't exist twenty years ago, and which most people today "speak" as badly as I spoke Hindi as a kid!

It was the world's dirty little secret: some of the time—*most* of the time—people couldn't make heads or tails of the tone behind messages they were getting in emails, text messages, conference calls, and so on. Nor were they entirely aware of how their own messages were being received. More than just a glitch or a nuisance—*technology is such a pain!*—our shiny new communication tools were causing serious issues. Work and decision-making had slowed. Teams were in disarray. Employees were left unmotivated, distrustful, uncertain, and paranoid.

It seemed that misunderstood "digital body language"—or rather, the lack of a set of universally agreed-upon rules—was creating big problems across the globe: in workplaces, communities, and even families. Everyone knew about these problems, but no one talked about them, except anecdotally. We had all grown up knowing how to read and write, some of us better than others (says the girl who remembers the day in school when, reading aloud from George Orwell's *Animal Farm*, she pronounced the word "peculiar" as "peck-you-liar," which her classmates never let her forget), but there was no instruction manual about how to read signals and cues in a digitized world.

Instead, people at work were squandering hours or even days in uncertainty, anxiety, and disquiet.

I was hardly a Jedi Master at all this either. I'd wasted entire mornings endlessly re-reading a single email, trying to figure out what an ellipsis or the single-word query *Thoughts?* meant. I'd heard about friendships imploding over a WhatsApp conversation. What about the "like" on Facebook or Instagram from a colleague who hadn't returned your two recent phone calls? (Did it signal "I'm sorry"? Was it a prelude to calling you back, a way of testing the friendship waters? Or was it a signal that from now on, you and that person would now be communicating exclusively via social media? What did it all *mean*? Something? Nothing?) What about the executive who signs off every email with *Thank you*—doesn't *that* show clarity? On the face of it, sure—so why does it come across to his colleagues as insincere and inauthentic?

I genuinely believe most people have good intentions. They just may not know how to *convey* those intentions.

How can we re-establish genuine trust and connection, no matter the distance? By creating a nuts-and-bolts rulebook for clear communications in the modern digital world. Communicating what we really mean today requires that we understand today's signals and cues at a granular level while developing a heightened sensitivity to words, nuance, subtext, humor, and punctuation—things we mostly think of as the field of operations for professional writers.

But if you think writing clearly is a niche or inessential skill, think again. When asked what the best investment professionals could make in their careers was, Julie Sweet, global CEO of Accenture, answered, "Develop excellent communication skills."[1] Sweet added that any employee, even a junior-level one, could significantly heighten their value by "articulately summariz[ing] a meeting . . . put[ting] together a presentation and [sending] emails that are really salient and to the point."[2] Much has been said about developing top-of-the-line presentation and public-speaking skills, but Julie Sweet has seen the future, one in which a supposedly "soft" skill—communicating well, especially in your writing—is a critical competitive advantage.

What does good digital body language look like in action? It means never assuming that our own digital habits (e.g., answering every email we get within 30 seconds, or never listening to our voicemails) are shared by everybody else. It means taking a few extra seconds to ask ourselves whether our sentences, words, or even punctuation might be misinterpreted. It means being hyperconscious of the signals and cues we send out, constantly checking in with ourselves, and learning along the way.

The book you're holding in your hands decodes the signals and cues of who gets heard, who gets credit, and what gets done in our ever-changing world. It will serve as a common-sense playbook that will help you understand how to communicate your ideas, negotiate relationships, speak your truth, and build trust and confidence with people very different from you. In the pages ahead, I will introduce simple strategies to help you and your teams understand each other and banish the confusion, frustration, and misunderstanding that arise from email, video, instant messaging, and even live meetings. My mission is to help you get closer to anyone—intellectually, emotionally, personally, professionally—and make you stand out as a trusted, straightforward leader, no matter the distance.

PART ONE

Digital Elements of Style

●

CHAPTER 1

What *Is* Digital Body Language?

They had been going out for three years when the fight, conducted exclusively via text message, occurred. The fight lasted for hours, back and forth, until at one point, frustrated and weary, Laura tapped out, *So r we thru?*

I guess so, Dave replied.

Laura was devastated. She called in sick to work the next day, and spent the next 24 hours mourning the loss of her relationship by meeting with friends, looking through old photos, and crying. The next night, Dave appeared on her doorstep. Laura, puffy-eyed, answered the door. "Did you forget about the dinner we planned a few days ago?" he said. "You said we were through," Laura said. "I meant we were through *arguing*," Dave said, "not through as in *you and me*."

Oh.

Most of us have had exchanges like this in our personal lives (though maybe not quite so dramatic)—communications so confusing and crowded with intimations that we spend an entire day trying to make sense of them.

Now take these same dynamics and transfer them to the average workplace.

Jack, a midlevel manager, gets an email from his boss. The last sentence—*That'll be fine.*—leaves him anxious. The period that punctuates it seems to dominate the screen, a black bead, a microbomb, lethal, suggestive, and—Jack would swear—disapproving. *Did I screw up?* Or is he merely over-thinking it? If he's *not*, how can he work for a boss who's so oblivious about the implications of a period?

Here's another: A positive, enthusiastic female boss head-quartered in New York is assigned to lead a remote team based in Dallas. One of its members, a young guy named Sam, flies to New York a few months later for his first face-to-face meeting with his new boss. After a good preliminary discussion, the boss asks, "So what were your first impressions of me?" Sam hesitates, then admits they weren't all that good. Almost all of his boss's communications were no frills and to the point, leading Sam to believe she was unfriendly, withholding, and probably cold. In person, though, she's the opposite. What made him feel that way? she asks. Sam had to confess it was because she didn't use abbreviations or exclamation marks. !!!!!!!!!

When punctuation and acronyms set us off into bouts of uncertainty, self-doubt, anxiety, anger, self-hatred, and mis-trust, we can be sure we're living in unmapped times.

· · · · · ·

I grew up reading—and re-reading—the books of Deborah Tannen. In 1990, Tannen, a professor of linguistics at George-town University, published her book *You Just Don't Under-*

stand: *Women and Men in Conversation*. I wasn't the only one; *everyone* seemed to be reading Tannen's book. An analysis of how we talk to one another using indirection, interruption, pauses, humor, and pacing, *You Just Don't Understand* dominated national conversations, spent four years on the *New York Times* bestseller list, and was translated into 30 languages.

None of us needs a linguistics degree to know that the ways we communicate meaning today are more confusing than ever. Why? Well, Tannen studied body language almost exclusively in face-to-face interactions. Her work was informed by linguistics, gender, and evolutionary biology, but also by what you and I convey whenever we cross our arms, look away, or blink. None of us, including Tannen, could have predicted that the majority of our connections would be virtual today. Contemporary communication relies more than ever on *how* we say something rather than on *what* we say. That is, our *digital body language*. When the internet came along, everyone was given a dais and a microphone, but no one was told how to use them. We all just picked things up as we went along. And the mistakes we've made along the way have had real consequences in business.

· ·

**You see, these days, we don't talk the talk or
even walk the talk. We *write* the talk.**

· ·

Texts, emails, instant messages, and video calls are ultimately visual forms of communication. What's more, each of us has different expectations and instincts about whether it's appropriate to send a text or an email, when to look in the camera during a video call, how long to wait before we write someone back, and how to write a digital thank-you or apology without seeming sloppy or insincere. Our word choices, response times, video meeting styles, email sign-offs, and even our email signa-tures create impressions that can either enhance or wreck our closest relationships in the workplace (not to mention in our personal lives).

Today, roughly 70 percent of all communication among teams is virtual. We send around 306 billion emails every day, with the average person sending 30 emails daily and fielding 96.[1] According to the *Journal of Personality and Social Psychology*, 50 percent of the time the "tone" of our emails is misinterpreted.[2] *Fifty percent!* Imagine saying "I love you" to your partner, but half the time their response is "Yeah, right." Have I felt that exact feeling with my husband, Rahul, after a text exchange? Not gonna lie—I'm guilty too!

More data: the *New York Times* reports that 43 percent of working Americans spend at least some time working re-motely,[3] a percentage that skyrocketed during the Covid-19 pandemic. Another study reported that 25 percent of respondents said they socialize more frequently online than in person.[4] A 2015 Pew survey found that 90 percent of cell phone owners "frequently" carry their phones with them, with 76 percent admitting they turn off their phones "rarely" or "never."[5] The average person spends nearly 116 minutes every day—that's about 2 hours—on social media, which over an average lifetime would add up to 5 years and 4 months.[6]

Psychologist and science journalist Daniel Goleman was the first to popularize the concept of "emotional intelligence," or EI, in 1990. Emotional intelligence refers to our ability to read other people's signals and respond to them appropriately while understanding and appreciating the world from others' perspectives.

Today, "emotional intelligence" and "empathy" have become buzzwords. They are discussed at roundtables. They are part of every mainstream education curriculum. They show up in value statements across every industry—from professional services to healthcare to technology. And they are trademark words in political campaigns and media conversations. Leaders have sold us on the idea that seeing situations clearly from others' perspectives can transform leadership styles, work cultures, and business strategies. Empathy, it seems, advances morale, triggers innovation, drives engagement and retention, and raises profits. Surely everyone can agree we need more empathy in the world.

Why, then, are we all facing a crisis of misunderstanding at work?

Well, a big problem is that reading emotion within the digital nature of the modern workplace is difficult. When the concept of emotional intelligence was popularized, the digital era was in its infancy. Email was a barely unwrapped toy. The very first smartphones were thick slabs and rarely appeared at meetings. Texting was what European teenagers did. And video calls were a foreign species. Today, many organizations and communities exist exclusively behind a screen. We've shifted the way we create connections and, consequently, how we work with our colleagues as well as our customers, community members, and audiences.

The loss of nonverbal body cues is among the most overlooked reasons why employees feel so disengaged from others. If used properly, and at scale, empathetic body language equals employee engagement. Disengagement happens not because people don't *want* to be empathetic but because with today's tools, they don't know *how*. Yes, a CEO can say, "My office door is always open" and tell everyone he's "accessible" and "approachable." But what if he's never actually *in* the office and the only way to communicate with him is to jump into his daily queue of 200-plus emails or Slack messages?

Most of today's workplaces, in fact, *minimize* the conditions necessary to foster and augment clear communication, leading to widespread distrust, resentment, and frustration. There is more physical distance between teams. There are fewer face-to-face interactions. There is virtually no body language to read. Plus, every few months, things seem to get *faster* (or maybe we're all just imagining it), leaving us no choice but to adapt to the newest normal. We become more thoughtless. We grow more accepting of distractions and interruptions, more indifferent to the needs and emotions of colleagues and workmates. This digital disconnect leads us to misinterpret, overlook, or ignore signals and cues, creating entirely *new* waves of organizational dysfunction.

The question is, *why?*

We are cue-less. It's worth repeating: nonverbal cues make up 60 to 80 percent of face-to-face communication.[7] Anthropologist Edward T. Hall called these signals and cues—posture, proximity, smiles, pauses, yawns, tone, facial expressions, eye contact, hand gestures, and volume—"the silent language."

> *How do we create connection when up to 70 percent of communication among teams takes place digitally?*

Our ability to care is compromised. Remember how a handshake after a job well done used to go a long way toward making you feel valued? Today, when team members work in different spheres, departments, offices, and countries, a handshake is impossible. One research study inserted very small delays into video calls to assess how colleagues judged one another. For delays of only 1.2 seconds, people were more likely to be rated as less attentive, friendly, and self-disciplined than if there was no delay.[8] Even on video chats, a frozen screen or a weird echo makes it hard for attendees to feel that their contributions are being heard and valued, leaving us with the question:

How do we show appreciation nowadays?

Our timing is off. When someone standing two feet away asks us a question, we respond instantly. We also know when a conversation has come to a natural end. But today, we are no longer obliged to respond to someone immediately (we have stuff to do!). At the same time, responding to employees' or clients' "urgent" texts five hours later may leave them feeling ignored and resentful.

How can we find the balance between busy inboxes and response times that convey respect?

Our screens have altered our traditional body language. When we glance down at our phones or answer "quick" texts during business meetings, one-on-ones, or lunch discussions, we tend to lose track of our surroundings. We wrap up meetings faster than we should, overlook our colleague's facial expressions, his smiles, or the fact he just put down his pen to listen more closely. In sales conversations, it's even easy to miss the lean-in.

How do we stop digital tools from interrupting even our face-to-face interactions?

Lastly, as we all know, **technology creates masks**. Today we all have the option of concealing what we really feel and think. Choosing to email or text can cloak uncomfortable feelings— but it can also create a whole lot of ambiguity and misunderstanding. For people who like to keep their thoughts and feelings to themselves, screens provide good camouflage. But that's not how leaders are built. Even when you're on a video conference, gazing back at your own face on the screen makes it harder for you to relax fully and naturally into the conversation.

How can we remain authentic and
connected when a screen divides us?

The answer is understanding the cues and signals that we're sending with our **digital body language** and tailoring them to create clear, precise messages.

....

What is *implicit* in body language now has to be *explicit* in our digital body language.

....

WHAT IS DIGITAL BODY LANGUAGE
(AND WHY IS IT IMPORTANT)?

Digital Body Language offers a systematic approach to understanding the signs of the digital world just as we interpret those of the physical world. It will identify and explain the evolving norms and cues of digital communications within organizations, and in doing so it will help to create a set of common expectations for communicating, regardless of distance. It's a little like a French-English dictionary, but in this case my mission is to translate in-person body language to such things as punctuation, video call first impressions, abbreviations, signatures, and the time it takes us to press *Send*.

By embedding a real understanding of digital body language into your organization or group, you can implement communication processes that can provide both the structure and the tools to support a silo-breaking, trust-filled environment. This skill, in turn, will lead to enormous efficiencies, as people will spend less time wondering about that period or (lack of) exclamation marks.

Below are a few practical examples to help you understand how traditional body language has been "translated" into digital body language.

- **Traditional Body Language** involves: a head tilted to one side, signaling that another person is listening attentively.
  Digital Body Language involves: "liking" a text. Praising another person's input in an email. Making a detailed comment verbally or in the chat box during a video call when someone expresses an idea, instead of just saying, "I agree."
- **Traditional Body Language** involves: stroking your chin or pausing for a few seconds, signaling that you're thinking about what's just been said.
  Digital Body Language involves: taking a few extra minutes to respond to a text, indicating respect for what it says. Writing a long or detailed response to an email that shows thought and focus. Pausing during a conference call to take in what's been said instead of blurting out the first thing that comes to mind.
- **Traditional Body Language** involves: smiling. (It's contagious. Our smile lights up the areas of the brain linked to happiness, which is why the people we smile at tend to smile back and/or feel a stronger sense of connection with us.)
  Digital Body Language involves: using exclamation points and emojis (well, within reason). Adding a simple *Have a great weekend* to the end of an email. Laughing during a video meeting.
- **Traditional Body Language** involves: nodding. Bobbing our heads up and down makes us seem both interested and approachable. Nods,

like smiles, are infectious, meaning that if we nod while we're speaking, others are more likely to go along with what we say.

Digital Body Language involves: responding promptly to a text, showing engagement by replying to an email with substantive comments. Writing *I completely agree with what you're saying* in the group chat during a Microsoft Teams meeting. Using a thumbs-up emoji in a video meeting.

The skill set that this book teaches will set you up to be your best—to deliver new ideas, speak truth to power, operate with confidence in fuzzy situations, and engage others in ways that complement your brand. It will restore emotional nuance to team engagement in ways that are clear, transparent, and long-lasting. With this book as your guide, you'll be able to create new norms of collaboration and behavior that reduce misunderstandings and enable you to lead with more clarity.

Not least, my mission in *Digital Body Language* is to help you stand out as a great communicator (and thereby a great leader).

In order to truly understand this new ideal of communication, we need to understand the four laws of digital body language: **Value Visibly, Communicate Carefully, Collaborate Confidently,** and **Trust Totally**.

We'll take them in order.

Value Visibly

The first law is based on the fact that the traditional signals and cues we use to show our appreciation for other people—a relieved smile, a handshake, a handwritten thank-you note—are either invisible in digital communication or take too much time and effort to implement. **Value Visibly** is about being attentive and aware of others, while also communicating that "I understand you" and "I appreciate you."

Valuing Visibly means we're always sensitive to other people's needs and schedules. Valuing Visibly means we understand that reading the emails in our inbox with care and attention is the new art of listening. When we Value Visibly, we're willing to sit with others' discomfort without feeling the need to fix or resolve it. Valuing Visibly means recognizing other people—and not being in a hurry about it either.

Inevitably, Valuing Visibly leads to greater levels of respect and trust. One time, I tried to schedule a phone call with a senior executive who had expressed interest in working with me. Over the next five months, she rescheduled three times. But she didn't just cancel the meetings—*she stood me up*. After the first no-show, I sent a follow-up email, and her assistant rescheduled our call. (There was no "I'm sorry" from the executive, or even a madcap excuse.) The second time she didn't show up, her assistant apologized and again rescheduled. The third time, there was nothing, just dead air. A few months later, this same executive emailed me, as if nothing had happened between us, for advice on how to join a club where I was a member. This time it was *my* turn not to reply. Could I *really* recommend her to my peers?

By "respect," I'm not talking about niceties or apologies. Respect means that others feel appropriately valued, included, or acknowledged. Respect means proofreading your email before sending it. Respect means honoring other people's time and schedules and *not* canceling meetings at the last second or delaying your response to an email so long that people have to chase you down. Respect means not using the mute button during a conference call to attend to five other things as someone is talking. Respect means writing clear subject lines in meeting invitations that explain exactly why you are requesting another person's time. (At a minimum, respect is about spelling that person's name right!)

Valuing Visibly also means acknowledging that solutions that may work in one context might not work in others. Imagine you've just pulled an all-nighter on a project, and your boss responds with *ty* or *tx*. It's not enough, right? In fact, it can be enraging. Now, imagine handing this same project to your boss in person, and getting a smile and a "thank you." You'd feel better. Valuing Visibly is about making the time and effort to communicate the equivalent of a smile or "thank you" across digital channels.

Communicate Carefully

The second law of digital body language, **Communicate Carefully,** involves making a continuous effort to minimize the risk of misunderstanding and misinterpretation by being as clear as possible in your words and digital body language. We Communicate Carefully when we establish unambiguous expectations and norms about which channels we use, what we include in our messages, and whom we include on the recipient list. We Communicate Carefully when we know why each person copied on the message is accountable, and who is responsible for the next steps.

Communicating Carefully goes a long way toward eliminating confusion. Communicating Carefully allows for a consistent understanding of each team member's requirements and needs, helping to streamline communication and reduce inefficiencies in teamwork. Finally, Communicating Carefully leads to alignment.

Has the following ever happened to you? You and your team have been working hard to initiate a great new idea. When it's ready to go, everyone is exhausted but exhilarated. It's time to reap the rewards for all that hard work! However, at this point, the company lawyers step in, ask a few questions, and either kill the project outright or redesign it enough so it's no longer recognizable.

Or what about the marketing team at a professional services firm that spends months on a new product offering only to realize that another operations team had already created the exact same thing a year before?

Or how about the team that can't agree on whether a project was a success or a failure because they never agreed on the metrics of success?

Regardless of the scenario, time is wasted, energy is sapped, and the workplace mood shifts from buoyant to discouraged. Why did no one realize this was an issue? Answer: there was no clear communication between silos. The legal team wasn't brought in until the last second. The compliance officers had no say during the planning stages. Customers were not listened to. The marketing and accounting leads never got together and said, *Here is what I want, here is what I think, let's get on the same page.*

Communicating Carefully means that people have to agree on whether or not a given project is necessary or in sync with the organization. Communicating Carefully means keeping employees and teams informed and up-to-date, and then checking in consistently to support their efforts. Who is working on what, and why? Personally, I've lost count of the number of times team members jump into projects without taking ten minutes to consider the principals involved, only to find out three months later that another team has been doing the *exact same work.*

But the biggest impediment to alignment is a lack of clarity. Communicating Carefully restores that alignment through digital body language signals and cues—from realizing that a "brief" message is not always a "clear" one, to eliminating tone-deaf language, to everything in between.

Collaborate Confidently

The third law, **Collaborate Confidently,** is about the freedom to take conscious risks while trusting that others will support your decisions.

Collaborating Confidently means managing the fear, uncertainty, and worry that define modern workplaces—and understanding that even when things get crazy, employees are there to support one another and work together to avoid failure.

Collaborating Confidently means empowering people to respond with care and patience instead of pressing them to respond to everything immediately in a 24/7 workplace.

Collaborating Confidently means prioritizing thoughtfulness while reducing groupthink behavior. What does that look like? Well, it might mean allowing the one remote member of a team to moderate a live meeting, creating a sense of inclusion and also reducing the bias we tend to have toward teammates who are physically present in the room. It might mean using the virtual chat tool in a video meeting to collect team opinions before calling on people with different ideas to speak up instead of listening to the loudest people who agree with one another first. It might mean designing a structure for work requests in email so that no one is left speculating wildly about the meaning of an exchange that ends with the letter *K*. It might even mean something as straightforward as making sure that team members always have what they need to move forward.

Collaborating Confidently decreases the chance of getting simultaneously trapped between being *over*attentive and *under*attentive, for example, when you obsess over minor stuff in one email while racing through others and overlooking important details.

Collaborating Confidently frees us up to overcome our habitual fears and uncertainties and move forward to *action*. It allows us to quit obsessing over *Did she really mean this?* and *Is he angry at me but just isn't saying so?* and *Are they giving me the runaround?* Instead, we assume the best intent from others, knowing that no one will succeed at someone else's cost, or by misdirection, or by forcing the upper hand.

Trust Totally

The fourth and final law of digital body language, **Trust Totally**, happens only after the first three laws have been implemented, yielding 360-degree engagement. The "Totally" part of Trust Totally is key, since it implies the highest levels of organizational faith, where people tell the truth, keep their word, and deliver on their commitments.

Trusting Totally means you have an open team culture, where everyone knows they are listened to, where everyone can always ask one another for help, and where everyone can grant favors whose returns may or may not be immediate. Once the first three laws of digital body language are in place, leading to Trusting Totally, congratulations! You've broken the stranglehold of fear and uncertainty in your organization and are on your way to assembling the always elusive perfect team.

Why? Because when we Trust Totally, we get the most out of people. By creating psychological safety on our teams (beginning with our leaders' own digital body language), our actions create trickle-down calm. Once trust exists, anything that helps support it is prioritized, and everything impeding or distracting from it is addressed and dealt with.

But let's be clear: Trust Totally doesn't mean we extend unconditional trust to *everyone*—especially people with whom we've had negative or unresolved experiences in the past. Instead, Trust Totally refers to a workplace environment where no one wastes time sweating the small stuff, where an ambiguously worded message or late-to-arrive response doesn't automatically give rise to fear, anxiety, or insecurity, and where we confidently assume everyone is on our side. This is a pretty big ask these days, but Trust Totally *works*.

Over the years I've worked with some truly overbearing people. I once worked for a boss who invaded my every waking moment. She began drilling my inbox with emails at 9:00 p.m., usually when I was in the supermarket, pushing around a cart at the end of a long workday. If I didn't reply within five minutes, the urgent texts would begin: *I need the report on the Chicago meeting! Did you finish? Where is it?* I would surrender my half-filled cart, run home, work late, get it off to her by midnight, and fall asleep. At 6:00 a.m., I would wake up to a new text, along the lines of *Let's chat this morning to review the report*.

Clearly this was one case where Trust Totally was impossible to implement. I did *not* feel Valued Visibly; my boss's messages were *not* Communicated Carefully; and we were decidedly *not* Collaborating Confidently. In the years since, I've found that when a foundation of Trust Totally exists, people are much more willing to say what they mean without fearing criticism or reprisal, and they may in fact come up with substantive improvements, even in tough situations like the one in which I found myself with my old boss. Trusting Totally can help transform a passive-aggressive or dominating colleague from an obnoxious pill to, believe it or not, a decent person.

Finally, Trusting Totally leads to empowerment. Yes, I know, *empower* and *empowerment* are so overused that they've become meaningless. So often leaders tell teams, "I want to empower you," but they aren't willing to give up even a sliver of control to allow other voices to contribute. So, the concept can't help but ring a little false. But in the context of Trust Totally, empowerment means giving people full ownership of their work, as well as the resources they need to finish it.

Empowerment means everyone feels safe to speak up, to introduce a controversial perspective, or to simply say, "This isn't working for me" without fearing they've created new enemies. Empowerment implies high levels of psychological safety, clear channels of information flow, candid discussions about how comfortable people are with failure, and clear ways forward that embed respect, alignment, and action across the workplace.

· · · · · ·

Digital Body Language is for people whose bosses and colleagues drone on and on about teamwork but never seem to do what's necessary to facilitate it. It's for anyone swamped with in-person meetings, conference calls, emails, texts, and social media platforms, those who have thrown up their hands and decided to just *set it* and *forget it*.

In the next few chapters, you'll read stories, learn strategies, and adopt commonsense rules that are designed to strengthen *any* workplace. You will learn about subtext, punctuation, pace, pausing, delay, the signals and cues of power and dominance, and the differences in digital body language across genders, generations, and cultures. Whether you lead a team or work alongside someone you can't understand, or you simply wonder why there's so little empathy around you, this book is for you. My goal is simple: to save you time, liberate you from fear and worry, and make seemingly indecipherable signals and cues as clear as a handshake, a head nod, an eye roll, a smile, or a ringing "Way to go!"

SO YOU WANT TO COMMUNICATE . . .

. . . TRUST:

- Traditional Body Language: keep your palms open; uncross your arms and legs; smile and nod.
- Digital Body Language: use language that is direct with clear subject lines; end emails with a friendly gesture (*Text me if you need anything! Hope this helps.*); never bcc anyone without warning; mirror the sender's use of emojis and/or informal punctuation.

. . . ENGAGEMENT:

- Traditional Body Language: lean in with your body as another person is talking; uncross your arms and legs; smile; nod; make direct eye contact.
- Digital Body Language: prioritize timely responses; send responses that answer all questions or statements in the previous message (not just one or two); send a simple *Got it!* or *Received* if the message doesn't merit a longer response; don't use the mute button as a license to multitask; use positive emojis like thumbs-up or smiley faces.

. . . EXCITEMENT:

- Traditional Body Language: speak quickly; raise your voice; express yourself physically by jumping up and down or tapping your fingers on your desk.
- Digital Body Language: use exclamation points and capitalization; prioritize quick response times; send multiple messages in a row without getting a response first; use positive emojis (smiley faces, thumbs-up, high fives).

. . . URGENCY:

- Traditional Body Language: raise your voice; speak quickly; point your finger (or make any other exaggerated gesture).
- Digital Body Language: use all caps paired with direct language or sentences that end in multiple exclamation marks; opt for a phone call or a meeting over a digital message; skip greetings; use formal closings, Reply All, or Cc to direct attention; issue the same message on multiple digital channels simultaneously.

●

CHAPTER 2

Why Are You So Stressed?

Navigating Power Plays and Anxiety

We all love to hate the horrible bosses we see in movies like *Wall Street, Office Space, Swimming with Sharks,* and *The Devil Wears Prada* (if you've never watched Meryl Streep shred an intern with a cold stare as she croons sardonically, "Please bore someone else with your questions," stream *The Devil Wears Prada* tonight).[1] Yet are these Hollywood archetypes really that much more exaggerated than some of the toxic work colleagues we've all experienced?

I have my own true tale of the peer from hell that I like to share at cocktail parties.

My first job after business school was on the trading floor of Lehman Brothers, before everything went bust there. The Lehman culture at the time was *shut-up-and-do-as-you're-told.* My colleague—I'll call her Harriet—was a young associate who had been on the team for several years. Among other things, I was responsible for making updates to a team project, and that involved getting information from . . . Harriet. Every time I needed to get an answer to something, I sent Harriet an email. More times than I could count, she cc'd my boss in her reply. It felt like a weird form of intimidation, as if she were alerting a hallway monitor to oversee my work. I began noticing that Harriet was excluding me from meetings too. When I confronted her about it, she said it was an oversight. So why did it keep happening? I finally realized that the reason I was being excluded was because it allowed her to frame *my* work as *hers* (I realized this only when I saw she used *I* instead of *We* on team projects in email exchanges).

Many of us are familiar with the common power plays in face-to-face encounters. We've all been there when a boss or older teammate pulls away physically, or turns toward another teammate in a meeting, or avoids eye contact, or raises a dismissive eyebrow, or stops smiling or making friendly gestures. Or maybe a teammate begins interrupting you in meetings, leaving you out, or rushing you along, signaling they're just too busy to chat.

Harriet's aggression was hard to miss, but in the digital sphere, power play behaviors can be harder to interpret. They might show up as cursory, one-word answers to emails, long delayed responses to simple questions, overly formal language, or a failure to reply at all.

It's really hard to be on the receiving end of this kind of power play, particularly because the ambiguity of digital communications allows for both misunderstanding (on the part of the less powerful) and gaslighting (on the part of the more powerful). This chapter unpacks the common anxiety-producing signals in digital body language and how to avoid paranoia and confusion with anyone.

WHAT ARE YOUR DIGITAL STRESSORS?

- Am I talking too much?
- Are other people trying to take credit for my work?
- What if they think my idea is stupid? Will they think less of me?
- Does the silence on the phone or video call have something to do with *me*?
- Am I making sense in this email?
- Will the recipient interpret this message the wrong way?

GRAPPLING WITH AMBIGUITY

Your manager reminds you of an upcoming deadline. Is he just being helpful, or is he showing off his hierarchical dominance? How can you tell the difference?

When dealing with a lack of clarity from someone else, here are two questions that can help you decide what to do next:

Who has more or less power in the relationship?
How much do we trust each other?

Power = Speed

Think about how fast you might respond to a request from your boss who has power over you. In this case, your quick response acknowledges that power. Hop to! Now, think about how fast you would respond to your secretary or junior report, who don't have as much power. We may prioritize speed, clarity, and substantive messages with our bosses and clients but deliver one-liners with no subject line to a junior report. Why? Because our bosses' higher power level usually incentivizes us to be more careful with our digital body language, especially as we struggle to prioritize time in a busy workday.

Trust Plays a Part Too

How—and what—we signal also depends on how much we trust the person we're communicating with. If you email a close colleague who has worked with you for years and the trust between you is high, he's likely to interpret a curt message as a signal that you're busy. But if the trust between you is low because of a turf war at work, he may interpret your brevity as a sign of resentment or anger. Trust goes much deeper too—variables like age, gender, culture, and race play key factors in whether we assume good intent in others' messages.

The best way to handle ambiguous messages is through what I call the Trust and Power Matrix, a tool that can guide which digital body language signals to keep in mind when juggling various levels of relationships in the workplace.

Review the visual of the matrix below. The y-axis indicates your level of power relative to the person you are communicating with. If you have more power (they are your subordinate), then you will look to the top half of the matrix. If you're communicating with your boss or a customer, then look to the lower half. The x-axis indicates level of trust. If you have

TRUST AND POWER MATRIX

UP

A | B

FAR ← → CLOSE

C | D

DOWN

a close, trusting relationship with the other person, look to the right side of the matrix. Otherwise, look to the left.

If you're in quadrant A (meaning you have more power and low trust), it's important that you show others why they are appreciated. Simple things like *Thank you for your message* or *I can't look at this now but I'll get back to you* go a long way in helping others manage expectations.

If you're in quadrant B (meaning you have more power but benefit from well-developed trust), you may become comfortable overusing brevity in your communications with this person. Be clear on deadlines and expectations and don't assume that others "get what you mean."

If you're in quadrant C (meaning you have less power and low levels of trust), prioritize quick, thoughtful responses to tasks and don't be afraid to ask for clarity. Your goal should be to increase the trust in the relationship. And if you are totally lost, find someone who can guide you on what to do.

Lastly, if you're in quadrant D (meaning you have less power and a very trusting relationship), don't drop your guard and let your messages and work get sloppy just because you're generally on the same page.

My clients have told me that this simple matrix has been incredibly useful in helping to navigate power-trust imbalances in a relationship. Use the Trust and Power Matrix as a tool to understand which digital body language signals matter most to improve your communications.

GOOD INTENTION, POOR RESULT

Intentions matter *a lot* in power dynamics, and digital body language has a funny habit of distorting them. Early in my career,

I emailed the CFO of a large organization who had recently offered to introduce me to a colleague. Intending to show how much I valued her time and what I imagined was her over-stuffed schedule, I emailed, *Just following up, I know you're really busy and wanted to check in on the connection to John.*

It backfired. She replied, *I would recommend that you never send an email starting with reminding someone that they're really busy.* (That was the last time I heard from her, by the way.)

I hadn't meant to disrespect her—just the opposite. In hindsight, I should have been more careful in my choice of words, as the gap in our power levels was high and the trust between us low.

What if she was on vacation or not busy and thought I was making her feel bad on purpose? I used the phrase "I know you're really busy" to counter my insecurity in having to follow up but I should've been direct.

My lesson was that I needed to be cautious in this kind of relationship and adapt my digital body language since there was a large power and trust gap. Sometimes people are having a bad day, or maybe they're hell-bent on misinterpreting you, or they want to flaunt their power and will misinterpret you no matter what. Try not to be that person, because the relationship will never be the same.

SAY WHAT YOU MEAN, MEAN WHAT YOU SAY

The widespread disconnection between *intention* and *interpretation* in the digital world is exacerbated by a phenomenon known as the *online disinhibition effect.* This occurs when we drop our guard, forego formalities, and express ourselves online in frank, uncensored ways we never would dream of doing in person. According to John Suler's article in *CyberPsychology & Behavior*, the online disinhibition effect arises from the "anonymity, invisibility, asynchronicity, introjection, dissociative imagination, and minimization of authority" typically fostered during virtual interactions.[2] When you and I interact in person, our social cues—facial expressions, tone of voice, gestures, etc.—act as behavioral curbs. Unless you're angry or have repeated yourself 20 times, you probably wouldn't say "Get this done now!" to a friend who looks like she's on the verge of tears.

When teams struggle to understand the intentions of their communications, power plays, animosity, and resentment usually follow, eroding trust and diminishing collaboration and innovative thinking.

Before we dive into analyzing some of the most common sources of digital anxiety that obstruct our goal of optimal clarity, remember one thing:

Always be impeccable with your own words.

In my teaching years at Harvard, I noticed one student who always, *always*, paused before he spoke. Whether he was answering a question or presenting to the class, you could see how much time and care he took to think before speaking. Then and now, it struck me as a critical leadership skill, one that made him stand out from the pack. Being impeccable with your words involves really listening to and understanding what someone else says before responding with thought and consideration—both online and off.

A marketing manager of a consumer-facing organization once sent a sarcastic, tone-deaf email to the CEO for approval. Instead of responding with his first reaction, which was *Get a GRIP!* the CEO took the time to answer with grace: *I want to let you know that sarcasm at this time is not helpful.* If someone sends you a passive-aggressive email, like *I assume you are completing this for me??* resist the urge to retort, *No, I'm a card-carrying idiot who has never completed anything in her whole life!* Instead, respond with facts and specifics: *I am working on this right now, and you will have it by the deadline discussed, Friday at 10:00 a.m.,* or, *Our plan said I would finish this on Wednesday and here is a draft. Can you let me know if you need anything else?*

By being impeccable in your own language, you not only shine a light on negative or ill-considered communication, you also show others by example the right way to respond.

I RECEIVED A CONFUSING EMAIL, NOW WHAT?

HOW TO DEAL WITH AMBIGUITY

- Ask yourself: Are you confused by the choice of medium, the tone, or the actual message? If it's the medium, switch to a different one. Sometimes a phone conversation really *is* better than email, just as email is a better forum for thoughtfulness and perspective than a text conversation. If tone is an issue, assume the other person's best intentions and respond with facts. If the problem is the message itself, request clarity.
- If the message remains stubbornly ambiguous, get a second opinion from a trusted advisor.
- Step up and admit that you need clarification. Ask the author to answer the following questions: What's the issue? What needs to be done? How can I help the most?

AM I THE PROBLEM?

HOW TO AVOID CREATING DIGITAL ANXIETY

When writing to others, always ask yourself the following questions:

- Is my message clear?
- Is there another way (or two or three ways) that the recipient might interpret my message?
- If my message is confusing, is there another medium and style I could use to convey it more clearly?
- If I have more power, am I unintentionally terse, vague, or rushed?

SINCE WHEN IS AN EMAIL A RORSCHACH TEST?

The Rorschach test, also known as the inkblot test, was invented by the Swiss psychiatrist Hermann Rorschach back in 1921. This psychological test asks the subject to evaluate a series of inkblots and report what shapes or images they see. Then the subject's perceptions are assessed to determine their thought processes, preoccupations, and personality. For example, when you look at one of the inkblots, do you see the wings of a bat or a butterfly? Two hands cupped in prayer? A demon wearing a cape? Ice cream melting on a sidewalk? The answers say next to nothing about the inkblot, but they reveal a lot about how you function emotionally.

At work, we come up against the equivalent of inkblots every day. Here's one:

Jane Robinson
Re:
To: Erica Dhawan

Why didn't you finish this? -Jane

At first glance, this is a straightforward bit of communication, probably written in a hurry. But what does Jane's email *really* mean? Is this just how she learned to send emails in business school, or is there something else going on, e.g., a digital power play?

I'll throw it back to *you*, Rorschach style, as we explore the four most common types of anxiety-provoking digital body language. In no particular order, they are:

* Brevity
* Passive-aggressiveness
* Slow Responses
* Formality

BREVITY

> what does this mean?????
>
> we need to talk
>
> CAN U SEND ME THAT TODAY

Brief? Yes. Have you placed a chill on my heart? Again, yes!

An early experience I had working at a large consulting firm taught me a lot about how stressful short, to-the-point communications can be. At the time, I thought I understood signals and cues pretty well, but I wasn't as good as I thought.

Living and working in New York, I was in almost daily communication with a British senior partner based in London. With 3,500 miles separating us, we'd never met in person, relying exclusively on email and phone conversation. As a younger associate, I was eager to prove myself, and the London partner seemed enthusiastic about working with me too. Unfortunately, 90 percent of the time, I had no idea what he wanted or even thought.

As the senior leader on the project we were working on, he naturally set the tone of our communication style. His emails were as terse as haikus. Is *Send a brief on this client* any different than *Fog rolls into shore. The clanging of the red buoy?* Mirroring his brevity, I'd answer, *Details please.* The most phone time I could get with him was 7–10 minutes in between his client meetings and airport travel, and the feedback I received during these snippets of conversation only confused me more. *Work on this more*, he would say, and a few days later, *Let's iterate*, though the collaboration implicit in that phrase never occurred. And not once did he tell me *what* needed more work, or iteration. As time went on, I felt that it was impossible for me to succeed. Worse, he was unhappy with my work—I knew that—but with no guidance or feedback longer than a few email lines, I couldn't address what was wrong.

I couldn't do the work I felt I was capable of since (a) I never got proper digital feedback, and (b) the power imbalance left me in no position to demand it. I was left with constant project-related anxiety that only ended when I left.

Brevity from the upper echelons of power isn't exactly uncommon. At Morgan Stanley, there was a running joke that the more senior you were, the fewer characters you needed to express your gratitude in a text or email. You started your career with *Thank you so much!* and after a promotion or two, this was cut down to *Thanks.* Another promotion produced *Thx* or even *TX.* One senior leader just wrote *T.* He was so important, rushed, and in demand his own mother couldn't expect any more keystrokes.

Senior leaders have a well-deserved reputation for sending sloppy texts and sloppier emails. Poor sentences, bad grammar, atrocious spelling—*we don't have time to care*! Brevity *can* make a person appear important, but it can also hurt your business. Getting a slapdash email means that the recipient has to spend time deciphering what it means, which causes delays and potentially leads to costly mistakes.

One executive whom I'll call Tom was renowned for both his carelessness and his brevity. Once, when a direct report sent him an email asking, *Tom, do you want us to move forward with this plan, or should we gather more information?* Tom replied, *yes.* Thanks, Tom, we'll move ahead on one, or both, or neither. Imagine how much time his team wasted debating how long they had to wait before someone pointed out to him he hadn't answered the question!

Employee engagement expert Dr. Jaclyn Kostner has this to say to execs about sloppiness: "You have to find the time; otherwise, you're not fit for the job and somebody else should be doing it. Or maybe you need to offload some responsibilities, because there's no excuse for sending people cryptic emails."[3] Leaders don't have to respond to *every* message, but when important work guidance is required, their communications should at least be *clear.* Imagine how much better my first draft of the work would have been if my haiku-speaking boss would've taken ten extra minutes to explain his goals?

As the recipients of cryptic messages, we overthink things in an effort to fill in missing words and absent meanings, prompting a lot of stress and confusion. I once had a client, I'll call her Janet, with whom I had a multiyear business relationship. Janet and I were planning for an upcoming event that was still a few months away. I sent her an agenda for us to review two days before our call. She replied with an email saying, *Yes, let's talk. Also need to discuss the budget.* My heart fell. I assumed she planned to say that her budget had dried up and I wouldn't be paid adequately for my work. I was furious, I barely slept that night, and I was still in a bad mood when I got on the phone.

"I forgot what I committed to pay you," Janet said at once. "Can you remind me so I can put it in my budget?"

I'd unnecessarily wasted time and energy preparing for the worst. My relationship with Janet was intact, but because of the power imbalance, I was prone to anxiety about our business arrangement, which served to waylay my focus from what I *really* needed to get done.

HOW DO I RESPOND TO CONFUSING MESSAGES?

Here are some things you can do if you receive unclear, brief messages:

- If it's a work request, ask clarifying questions such as *Can you share what you need from me?* or *Thanks for this. When do you need this by?*
- If you're not sure about something, ask for the details you need to get a better idea of the other person's intent, as well as the task at hand.
- Change the channel of communication to a phone call, video chat, or face-to-face meeting for additional context.

If you consistently feel like there's a disconnect between your messages and the responses you're getting:

- Ask yourself: Is it clear what the recipient needs to do, why they need to do it, and when they need to do it by?
- Ask yourself: Am I using the right communication channel? Would a quick phone call provide more context than an email?

PASSIVE-AGGRESSIVENESS

As per my previous email,

I'll take it from here . . .

Am I missing something?????

We've all felt it. That moment when we're called upon to interpret phrases that *could be* perfectly fine but tie our stomachs into knots anyway. What does she *really* mean in her text when she types, *Per my last email,* or, *Just a gentle reminder . . .*

She sounds as wise and gentle as a Norse goddess—but is she actually saying, "You didn't read what I wrote. Pay more attention, goddammit!" or, "Get this done! It's late! I'm waiting!"?

Sometimes we perceive coded language as a microaggression, one that fuels already bad feelings among co-workers. Other times, we tell ourselves it's probably just a phrase our boss picked up in business school and doesn't realize how patronizing and stuffy it looks in print.

Here's an example: Melissa and Rosalee were co-workers who hit it off immediately. But when they started working on the same project together, things started going downhill.

Consider their following GChat conversation:

Melissa: Hey girl! I know you're busy, but could you get me that draft report today?

Rosalee: Oh hey! Yeah of course, amiga. Technically I have until tomorrow, but sure, whatever you might need!

Melissa: Thank you soooo much! Actually the project schedule shows it for yesterday, but I didn't want to bug you then, just to make sure we're on the same page going forward I'll send you a link to the schedule 😃😃😃 thanks for sending that report over right away, want to grab coffee soon?

Rosalee: Thanks for the offer, but no, actually have to work on that report now. Don't worry though, I'll make sure to review that schedule too, right away. Enjoy your day Melissa.😎👍🏿👍🏿👍🏿

I'm guessing those two won't be sharing a meal together anytime soon. Instead of softening the conversation with her request about lunch, Melissa should have been more direct from the start, getting to the point clearly while avoiding vague language like *I know you're busy, but* and *just to make sure we're on the same page.* Melissa *could* have written: *Hey! My calendar says you're set to finish the report today. Can you let me know when I should expect the document?* A very simple message showing the source of her information (her calendar), along with a clear, direct ask.

On Twitter, Washington, D.C.–based writer and marketer Danielle René posted about the subtle ways people use everyday digital language to denigrate one another.[4] *Per my last email* was the top choice for subtly correcting or even shaming a sender. René also asked her Twitter followers to send in their best double-edged hostilities, a request that soon went viral (10,000 retweets, 40,000 likes, more than 1,000 responses).[5] The clear winner to my mind was an email with the subject line *Friendly Reminder* and the words *I wanted to bump my previous message to the top of your inbox as I know you have been really busy.*

Among the highlights of René's Twitter survey:

@chocolateelixir: I love forwarding previous emails and saying "correct me if im wrong but here you stated . . ."

@darkandluuney: "Just to reiterate . . ." and then highlight and bold what was clearly stated in said email chain.

@crumr018: "not sure if my email made it to you as I haven't heard back"

@_verytrue: I LOVE an "Any updates on this?" (The sender went so far as to attach an email to her email!)

In all of these cases, the recipients may not understand the subtext of what's really being said, but is that really the point? By deriving gratification from writing, *Just to reiterate . . .* and *Any updates on this?* the *senders* feel better. Petty? Sure.

For better or worse, digital communications don't let us see each other's immediate reactions—which is why we look for ways to "politely" express irritation. The key word is "politely." While some of these phrases can be construed as passive-aggressive, the truth is that busy people (especially older ones) often use them as legitimate follow-up requests, no passive-aggressiveness implied.

I struggle with one phrase used by more than one of my clients. *Thanks for your patience.* Whenever I see *Thanks for your patience* in an email, I can't decide if they're brushing me off with an undefined future date or if they really only need a few days longer than expected to get back to me. I know, I know—in most cases they're saying, "Sorry I'm late with this, it's taking longer than I thought." That's all. No need to lose sleep over it.

Passive-Aggressive Feelings Behind Common Phrases	
Per my last email	You didn't really read what I wrote. Pay attention this time!
For future reference	Let me correct your blatant "mistake" that you already knew was wrong.
Bumping this to the top of your inbox	You're my boss, this is the third time I've asked you. I need you to get this sh*t done.
Just to be sure we're on the same page	I'm going to cover my ass here and make sure that everyone who refers to this email in the future knows that I was right all along.
Going forward	Do not ever do that again.

So how should we frame our own *just following up on this* without engaging in any passive-aggressiveness ourselves? When is it considered okay to loop in our boss without seeming like a jerk? When do we text a response rather than emailing it? When do we use the phone to call and clarify something?

TRUST AND POWER MATRIX

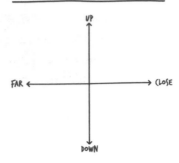

Well, let's go back to the Trust and Power Matrix.

Who between us has more power? How much do you and I trust each other?

If you have a high level of trust, opt for the phone call, and don't hesitate to respond quickly and informally. If you have less trust or a higher gap in power levels, be specific and polite in your responses and use formal channels. (There is often a fine line between polite-sounding and passive-aggressive messages.)

HOW DO I DEAL WITH MY PASSIVE-AGGRESSIVE CO-WORKER OR BOSS?

- **Avoid responding to messages or emails when you're angry or frustrated.** This prevents miscommunication, wasted time, and regret. If you feel emotionally hijacked, save your email message as a draft and revise and send it when you're in a better mood.
- **Stay in the place of reason.** Think through your responses and give people exactly what they need to take action. Assume good intent. Step into their shoes and ask, "Why might they have made a mistake like this?" Sometimes simply adding a quick brief to the recipient so he or she doesn't have to read previous emails (*Here's what I need from you* or *Here are the open dates*) is helpful.

- **Show empathy and encouragement.** Replace imperatives like *Do this* with conditional phrases like *Could you do this?* When delivering feedback, begin your message by expressing appreciation, using words like *Thank you* or *Excellent job on . . .*

SLOW RESPONSES: HOW. COME. YOU. DON'T. CALL. ME. ANYMORE.

If your normally happy colleague ignored your chirpy "Hello!" in the hallway, you'd know something was wrong. If the silent treatment continued when you were back at your desk, you'd try to figure out what happened. It's not so much your colleague's silence that makes you so anxious, it's the change in behavioral patterns.

In our digital world, the so-called silent treatment can show up as delayed emails, texts, and even ghosting behavior, which in turn triggers a phenomenon I call "timing anxiety"— the intense worry we feel when we find ourselves ruminating over the potential meanings of digital response times. Timing anxiety can last hours, days, weeks. Was the other person just . . . busy? Did your email even arrive? Did it end up in a spam folder? Or is the person not returning your message on purpose and engaging in what I call *the silent response?*

THE SIGNS OF SILENCE

Sometimes we get a response to an email that's so lacking in expression or emotions it might as well be a fast-food flyer slipped under a lobby door. If this happens, it's impossible for us not to wonder, *Am I overreacting?* Is it possible the other person is just being direct and to the point?

Our collective and universal reliance on fast-paced, real-time texting often leaves us unnaturally frustrated when we don't receive instantaneous responses in other channels. Imagine you just emailed a co-worker on a different team with the words *Dinner soon?* Two days later, no response. But your co-worker *has* found time to post a new picture of his corgi on Facebook and Instagram. Instead of following up with a new email, you "like" his social media corgi photo, hoping your little red heart will guilt him into responding to your first email. A week goes by before he responds with *Sorry for the delay!!!!* When you finally meet, it turns out that he really *was* just overwhelmed with stressors (training a corgi puppy takes work) and didn't have the emotional bandwidth for a dinner meeting. Remember the days of voicemail, when a one-week response was acceptable? Here's an illustration of the emotional rollercoaster we go on with slow response times.

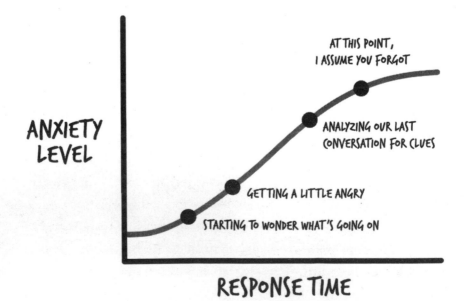

ANXIETY LEVEL

AT THIS POINT, I ASSUME YOU FORGOT

ANALYZING OUR LAST CONVERSATION FOR CLUES

GETTING A LITTLE ANGRY

STARTING TO WONDER WHAT'S GOING ON

RESPONSE TIME

Other scenarios are just as ambiguous. My friend Margaret told me that when she left one company for another, a former colleague stopped talking to her. Margaret texted her colleague to tell her she was leaving, and waited eight days before receiving a response. In Margaret's mind, a two-day delay was tantamount to not talking at all. Another friend, Julie, told me that when someone took a week to reply to her urgent text, she was so annoyed she didn't want to answer. Interpreting the weeklong silence as being "ignored," she ignored the person right back. Unfortunately, there are no hard-and-fast rules to know definitively if someone is using silence as a cudgel. The bigger point is that we all need to be aware that our digital body language emits signals, deliberate or not.

Ghosting is a relatively new term used to describe the act of allowing texts or emails to go unanswered—especially when a follow-up has been sent to no avail. A few months ago, another friend of mine, Neill, sent his friend Shelly a WhatsApp message: *Can you call me when you get this?* The notification popped up on Shelly's phone, but she was annoyed with Neill at the time and in no mood to answer, so she devised a workaround to make it look like she hadn't read Neill's message by sliding it down and reading only the message's sneak preview. Seeing that Shelly never actually opened his message, Neill indeed assumed that she hadn't read it. For her part, Shelly felt in the clear so long as she refrained from clicking on it. When enough time had passed, Shelly finally opened the message on WhatsApp and replied, *Oh, hey, just getting this! I'll call you in a sec.* (They both told me later this is how their exchange played out.)

Because of the expectation for immediate responses, today's messaging systems make it nearly impossible to take a break from one another. We've all felt like Shelly or Neill at one time or another. We can't assume our propensity for fast (or slow) responses is shared by others. But in our workplaces, it's essential to establish agreed-upon norms around messaging apps and time frames so we don't end up "ghosting" one another over, say, a simple communication about an upcoming meeting.

AVOID DIGITAL GHOSTING

If you are waiting for a response:

- Don't jump to conclusions. Unless it's critical that you get a reply ASAP, remember that people may have a lot on their plates.
- If you follow up twice with no response, switch to a different medium.

If you need to get back to someone:

- If you can answer in 60 seconds or less, respond immediately.
- If it's urgent, respond immediately or let the sender know you are *working on it*. Make an appointment with yourself on your calendar to answer.
- For matters lacking urgency, don't stress. Block out time to follow up later at your convenience.

FORMALITY: STANDING ON CEREMONY

The world of work is inarguably less formal than it was even a decade ago. From the way we dress to the way we interact with those above and below us on the org chart, times sure have changed. Even my corporate law or big-four accounting clients agree that formality has generally gone down in the workplace. For this reason, slight increases in formality levels can feel like unspoken power plays. Sometimes being too formal can make you seem unfriendly or standoffish—putting you at odds with others.

"Thank you" is a baseline etiquette favorite for good reason. *Thank you for dinner. Thank you for returning my call so quickly.*

Thank you for your time. But what about colleagues, especially peers, who use those two words to buttress their own professional power? *I'll need the report by 5:00. Thank you. I will be ready at 8:00 a.m. Thank you.* When "thank you" is used like this, it goes from being a simple statement of gratitude to a royal fiat—and needless to say, it's hard not to bristle when a colleague sounds like she's throwing her crown and scepter around.

When people you work with shift from friendly to formal, it can be disquieting. For example, when your boss begins an email with *Dear Steve*, instead of what he usually does (e.g., plunge in without even mentioning your name), what are you, Steve, supposed to make of it? What about if you see *Best* instead of *Thanks!* on an email signature from a longtime colleague?

Trina, for example, is known for her informal communication style within her team. One day, she caught them off guard by sending out a long email listing a number of work requests, each one bolded in a different color. Proud of her clarity and directness, Trina was certain it would lead to good results. Confirming receipt of the info while also trying to be funny, one team member, Dianne, replied to the whole team with a lighthearted comment: *Wow, what a colorful email! ;) We're quite the team to wrangle!* Trina replied, *Well, I am the manager here . . .* Uh-oh. Trina felt that Dianne had forgotten her place, and she signaled *her* displeasure by pulling rank. While this tone *did* succeed in reminding Dianne that Trina called the shots, it also did a number on their relationship.

SIGNED, SEALED, DELIVERED

Greetings and signatures also imply levels of emotion depending on their formality or informality. If you typically sign off with *Sincerely* or *Respectfully*, odds are you prefer to keep recipients at a polite arm's length. It's good to be authentic if you are naturally more formal. But if you are trying to build close friendships at work, this kind of formality may cost you.

This range in tone extends to our job descriptions too. A former executive at Bank of America once described his time there: "Whenever I wanted to get a response from someone in the organization who didn't know me, I always added my formal title in my email signature that said I was a VP. I always got quicker responses." Alternately, if you begin emails with *Hey*, or end a one-liner email with a smiley face, recipients can be assured that you're a fairly informal person.

Even the pronouns we use in our emails signal the level of formality we prefer, not to mention the power dynamics within a relationship. Psychologist James Pennebaker found that "[i]n any interaction, the person with the higher status uses I-words less (yes, less) than people who are low in status."[6] Pennebaker tested this theory with his own correspondence by analyzing his email exchanges at the university where he taught. Of his findings, he reported, "I always assumed that I was a warm, egalitarian kind of guy who treated people pretty much the same. When undergraduates wrote me, their emails were littered with I, me, and my. My response, although quite friendly, was remarkably detached—hardly an I-word graced the page. And then I analyzed my emails to the dean of my college. My emails looked like an I-word salad; his emails back to me were practically I-word free."[7]

To reduce anxiety among team members while creating greater transparency, leaders should create clear guidelines around the use of cc, Reply All, titles, and any other signals used to denote hierarchy. Mike, the president of a large technology company, told his employees that any message on which he was cc'd ended up in a separate email folder that he checked weekly and to which one shouldn't expect a response. Across the board, Mike's employees were more intentional than most teams I've encountered. Thanks to clear norms, they knew when to use each box and what to do if they found themselves enmeshed in a never-ending email chain.

HOW FORMAL SHOULD I BE IN MY DIGITAL BODY LANGUAGE?

- If it's a new relationship, follow the formality level of the person who has more power.
- If it's a longtime trusted relationship and the formality changes (suddenly or gradually), ask yourself why, or consider checking in with the other person.
- If it's a longtime relationship with an obvious power gap and the formality changes, take the lead of the more powerful person and mirror that change.

As with any form of communication, you will do well to assume the best intent when interpreting other people's digital body language. Don't automatically conclude that you are being singled out just because someone else's digital behavior is confusing. Give other people the benefit of the doubt: maybe they were in a hurry, up against a deadline, or (more than likely) oblivious. It happens to the best of us. Often we are the last to know if we have offended someone unwittingly.

If it still bothers you, reach out to the person by phone or video chat if not in person. Explain the source of your anxiety openly, without being overly apologetic or accusatory, and ask for clarification. Doing so will help you build trust and connection, no matter the distance.

•

CHAPTER 3

What Are You *Really* Saying?

How to Read Between the Lines

When I was growing up, my mother told me over and over again, "Erica, stand up straight!" I would try, but slouching always felt more comfortable to me than standing tall. It wasn't just my mom either—my teachers also admonished me for my poor posture. But it wasn't until I got my first job that I straightened up forever. "Your posture signals your confidence," one of my mentors told me, adding, "Slouching makes you look unprofessional." Having always associated slouching with my own laziness, I realized it was irrelevant how slouching felt to me. It mattered what it communicated to others *about* me.

Think about the last time you were in a meeting. Who was at the head of the table? Who showed up a few minutes late? Who made it a point to sit next to whom? Whose chair was pushed back ever so casually from the table? Who spent the entire meeting checking email? These are among the signals and cues that indirectly illustrate confidence, influence, and power at work.

Now: Would you be able to identify those same signals and cues during your last conference call, video meeting, group text, or Slack channel? Direct IM messages and email discussions during conference calls have taken the place of distracting side-bar conversations during physical meetings. Table seating politics now show up via To, Cc, or Bcc lines, and the order in which your name appears (is it first? Last? Squeezed somewhere in the middle?). Offline, direct eye contact has the power to say *a lot*. Today, a terse email written in a clean font with a period at the end of the last sentence has the power to intimidate from thousands of miles away. The enthusiasm and nodding heads of in-person meetings now show up as exclamation marks, emojis, and rapid responses. Entrances, exits, and goodbye handshakes used to tell you how a meeting went. These days, based on the tone of the greeting and the sign-off in colleagues' follow-up emails, you can only guess how they feel that same meeting went.

Digital body language may be fundamentally casual, but *casual* isn't the same as *careless*. In all good communication, every word and signal counts, especially in an era when we can no longer rely on the sound and tone of our voices, and where eye contact is absent.

In this chapter, I'll explore how real-world body language translates onscreen into words, punctuation, timing, and choice of medium; the countless ways the words you type can be misinterpreted by the receiver; and how to gain control over those words before you click *Send*.

Below, you'll find the most important digital body language signals we send out every day, and what they correspond to in real life:

- Priority = Choice of Medium
- Emotion = Punctuation and Symbols

- Respect = Timing
- Inclusion = To, Cc, Bcc, Reply All
- Identity = Your Digital Persona

DIGITAL MEDIUM CHOICE— THE NEW MEASURE OF PRIORITY

Choosing the best medium to use—email, Slack, phone, or text—is essential and depends on context. First, how important or urgent is your message? Second, to whom are you communicating? Do you want to tell your colleague about a minor typo in the presentation she's about to give? If so, what's better—email, Slack, the phone, or a text? What if instead of your colleague, it's your boss?

Effective though each may be, every communication channel brings with it a set of underlying meanings and subtexts. Learning how to navigate this confusing array of hidden meanings is a telltale mark of digital savviness and—ultimately— professionalism.

Example: as the newly appointed CEO of a large organization, Adriel needed information about an especially difficult client. She sent a late-night calendar invitation to Brian, the account manager, for early the next morning.

It looked like this:

New Meeting: (No Subject)
8:00 a.m. - 9:00 a.m. Friday
From: adriel@doe.com
To: brian@doe.com

Brian was there at the scheduled time, though he seemed unaccountably anxious. Once he and Adriel began talking, the relief in his body language was palpable. When she said as much,

Brian blurted out, "When I got the invite last night, I couldn't sleep. I thought I was getting fired." His boss was surprised. To Adriel, a calendar invite was and is nothing more or less than a neutral statement: *Make time for your boss. Here's the date.* But to Brian, receiving a calendar invite without context felt so cold and impersonal it could mean only one thing: he was no longer worthy of even the most standard office niceties and was on his way out the door, probably in handcuffs. Not knowing how to read Adriel's choice of medium, Brian assumed the worst.

CHANNEL-SURFING

Switching between channels can indicate a shift in the urgency of the message—or even denote the closeness of a relationship. When I was growing up, my dad, a physician, always carried around his pager. Whenever it beeped, my siblings and I knew he had to excuse himself, go into his office, and call the hospital to attend to one of his patients. Back then, I had my own instrument, my own private code of urgency, in the form of AOL Instant Messenger. If my best friend IM'd me *911*, I was supposed to call her immediately from my house phone to get the latest hot gossip.

When we make the time to text, call, or even stop by a colleague's office to say "Way to go!" we're communicating connection and respect. But if a boss emails positive feedback to a project, leaving out the "Way to go!" part and proceeding to pepper the employee with multiple texts questioning this, that, and the other thing, the employee might wonder what's going on. What's with all this weird, undivided attention? Is something *wrong* with the project? Why is the boss communicating in such a rapid-fire way that the employee has no time to look things up or respond? When bosses have seniority—and they wouldn't be bosses if they didn't—everything they do or say carries transferential weight, so we place even *more* urgency on a change prompted by the boss.

That said, channel-switching can be used to anyone's advantage. Any boss (or employee) can regain control of inbound requests by switching over to the medium of their choice. For example, if you just received a text but want to slow things down and think through your answer, responding via email sends the unmistakable message that the issue warrants a more thoughtful approach. Alternately, if you're the person trying to reach someone, you can forge a better connection (and probably get a more satisfying response) by sending a request on *their* preferred medium.

Finally, there are those times when you realize you've chosen the wrong channel, period. For example, three responses into an email chain, you might realize that the particulars of your conversation are too complex to discuss asynchronously and that scheduling a video call or live meeting is a much better way to proceed.

(DON'T) CALL ME!

A few years ago, Alisa and I, colleagues in the same industry, were getting to know each other, mostly via email. At one point, we made plans for a Saturday night dinner. Later that week, I realized I needed to postpone and reschedule. It was the third time in three weeks I'd canceled, and I felt terrible. I didn't want to seem flaky by sending Alisa yet another email, so I called her, and when she didn't pick up, I left a voicemail. Two hours later, Alisa texted me: *Did something happen?* When I told her I needed to cancel our Saturday dinner, she was relieved. Later I learned she'd assumed something catastrophic had happened. After all, why else would I have switched mediums so abruptly?

This mad whiplash of emotions is familiar to most of us. More disconcerting than what is said (or not said) is the change in pattern of *how* or *when* it's said. Oftentimes, we assume the worst when faced with unexpected digital body language, when most of the time it's nothing at all. *I was misinterpreting the signals. I was just anxious.*

When Alisa and I finally met up in person, she confessed to "phone-phobia." As a 40-something female, she'd gotten so accustomed to colleagues and friends communicating via text and email that she would get flustered, even panicked, when her phone would ring out of the blue. She's not alone. Many of us (who grew up with IMs and texts as our primary form of communication) are so used to controlling when and how we respond to texts and emails that when a phone call comes in, we treat it like a bomb that's set to detonate on the sidewalk. We feel vulnerable, unready, and even invaded, especially if we haven't already established a relationship with another person via email.

We all have our preferences for certain mediums—texting, of course, is universally popular—and a distaste for others, say, Zooming, or talking on the phone. Sarah, a 25-year-old who works at an advertising agency, once vented her frustrations about her boss's choice of medium: "Every time I email my boss a report, he calls me back with his comments and questions instead of just emailing me back! Ugh!" Her exasperation is similar to Alisa's phone-phobia, which involves her "feel[ing] caught off guard or put on the spot" if she gets an unscheduled, out-of-the-blue call.

That said, some people feel that a phone call is more efficient, personal, and collaborative than an email. It is, but with a caveat. While phone and video calls are a popular medium, they also give rise to rampant multitasking, not to mention the fact that the awkward pauses endemic to these mediums often make us perceive people differently, especially if we're meeting them for the first time.[1]

> "Hey! Can you hear me?"
>
> "... What? ... Oh, yeah I can hear you! Hi!"
>
> "I said can ... oh great! Okay then, let's dive—"
>
> "What's great?"
>
> "[siiiigh] ... I think we have a delay on the line ..."
>
> "... Yeah ..."

I CAN'T HEAR YOU!

I'm old—or young—enough to remember a series of Verizon TV commercials from the early 2000s where viewers saw a spokesperson clutching a phone and standing in a cornfield, then on a boat on the Hudson River, then in a children's playground, and finally on top of a snowy mountain. "Can you hear me now?" he keeps bellowing. Well, if his voice was audible on a mountaintop in 2001 using 2G, why can't I make out what other people are saying during a Zoom call in my living room? It turns out that technology always has its limits.

Mita Mallick, former head of diversity and inclusion at Unilever and now at Carta, once tried to express an opinion during a Zoom video meeting attended by 25 colleagues. "I'm interrupted, like, three times and then I try to speak again and then two other people are speaking at the same time interrupting each other," Mallick said.[2] When she finally managed to get in a word, she couldn't gauge anyone's response. The same thing happened when she cracked a joke—was anybody laughing? Did people agree with the points she was making? What did all those blank stares *mean*?

The delays we experience on technologies like Microsoft Teams and Zoom (and even during some phone calls) add up. We pause between sentences in anticipation of an encouraging nod from our colleagues, and when one doesn't come, the silence can feel literally unbearable. *Don't* pause, and we risk interrupting someone else without meaning to. Pause *too* long, and everyone on the call grows quiet. Along with wasting everybody's time, delays alter how we perceive one another. Worse, the mechanics of video calls mean that we can look either at our screens or our cameras, but not both, making direct eye contact impossible.

Here's some advice—before any Zoom or Webex session, acknowledge the obvious: the video call format is inherently awkward. It's no one's fault. It's just the technology. To minimize the awkwardness, request that everyone have their cameras on, that people signal their desire to speak with the hand-raise feature, and that backgrounds be kept distraction-free. Another tip? Take control of the pause. Once you're done speaking, ask if everyone understood what you were saying. Are there any questions? If there are, add them to the group chat, and wait two minutes before you begin to address them.[3] This camouflages the delay while the participants process and formulate their questions.

A team's choice of channel will vary based on the organizational culture, but all leaders should set sensible, easy-to-follow norms for their teams. Generally, that means that whoever is in charge must take the time to analyze the team's use of digital mediums, ask the team where they experience misunderstanding or awkwardness the most, and clarify a path forward. A great example of this kind of outline can be found on page 124.

Within each medium, it's important to respect boundaries. Non-emergencies can generally wait until work hours. If a last-minute email about a meeting scheduled for the next morning is sent outside of reasonable work hours (7:00 a.m.–7:00 p.m.), it probably warrants a text message too. Avoid sending a dozen instant messages in a row if you haven't gotten a response to the first one or two. Choose email for directional updates or work requests (this also ensures that there are records of your communication). Phone calls or video meetings are best for deeper collaborative discussions and decision-making.

Of course, there are exceptions to these boundary rules. You may shoot off a quick 10:00 p.m. text when the start time for tomorrow's client meeting suddenly changes. But if you are going to cross the boundaries, you should have an extremely good reason to do so, and should not make it a habit. Habitual boundary-crossing is the fastest road to team burnout.

SHOULD I EMAIL, TEXT, OR SCHEDULE A CALL?

Before choosing a channel, use the following questions to guide you:

Am I trying to have a quick conversation?
Does my message cover a lot of detail?
How fast do I need an answer?
How formal or informal is my relationship with the recipient?

(We will go deeper into what factors will influence your choice of medium in chapter 5: "Communicate Carefully.")

PUNCTUATION AND SYMBOLS—
THE NEW MEASURE OF EMOTION

You receive a text from a colleague: *Just sent plan to Jason!* Huh? What? You type, retype, and change your mind a half-dozen times, and you *still* don't know how to reply. Why are you stressing out over something so inconsequential? Because each of the following four options carries a different—and infinitesimally subtle—meaning.

As I said in my introduction, nonverbal cues (facial expressions, gestures, tone of voice, pitch) comprise nearly three-quarters of how we understand one another in person. As we know, our computer screens filter out these and other signals and cues, stripping away many of the qualities that make us human, forcing us to adapt the emotional logic—if it exists—of computers.

By way of compensation, our language has become a lot more informal. In an effort to infuse our texts with tone while simultaneously guarding against possible misinterpretation, we might type *I'm so so soooo SORRY!!!!!!* instead of plain old *Sorry* after missing a conference call or canceling a lunch date at the last minute. To clarify our feelings further, we roll out symbolic content ranging from emojis and hashtags to "likes" and LOLs. But instead of clarity, most of us come up against *more* confusion.

WHEN YOU'RE SMILING

Brody and Jessica are both relatively new employees working for the same company. Brody came from a startup; Jessica had a background in Big Law. They recently were tasked to collaborate on a project. It didn't take Jessica long to get annoyed with her new colleague.

Why? Well, Brody had a habit of sending brief emails stocked with emojis and abbreviations. Jessica emailed him back using a formal, professional style—no abbreviations, no emojis—hoping he would get the message that he was being too informal. He did get the message, but not as she intended it. He decided that Jessica was uptight—no fun, a faceless, imperious queen bee. Brody, by contrast, was raised on a diet of hearts, smiley faces, and enthusiastic punctuation. He saw those things as signals of approachability, friendliness, and camaraderie. Jessica simply saw them as rude and assumptive. Brody and Jessica continued to vex each other until the project ended, at which point they started trash-talking each other across the organization.

Neither Brody nor Jessica was right or wrong. They each had their own communication styles. But with all respect to Jessica, emojis and punctuation marks *are* useful tools to infuse emotion into otherwise flat, one-dimensional digital communications. Even on Zoom or Webex calls, tools like chat or the thumbs-up button convey our energy and even humanity.

Consider your own use of punctuation marks and symbols. How do you want your relationship with the person you're messaging to progress? If it's formality you're after, and you feel uncomfortable using excessive signals, then stick to the bare-bones facts and end your sentences with periods. Also, if your boss's or client's digital body language is formal, my advice is to mirror that formality back. If, on the other hand, you want to build closeness, and the other person seems receptive, go ahead and dial up the smiley faces and LMAOs.

I'M SO EXCITED!!!!!!!!!!

In one of my favorite *Seinfeld* episodes, Elaine, a book editor, comes home one day to find that her boyfriend, who's also one of her authors, has written down her phone messages on a pad of paper. The following dialogue ensues:

> Elaine: I was just curious why you didn't use an
> exclamation point?
> Jake: What are you talking about?
> Elaine: See, right here you wrote, "Myra had the
> baby," but you didn't use an exclamation point. I mean,
> if one of your close friends had a baby and I left you a
> message about it, I would use an exclamation point.
> Jake: Well, maybe I don't use my exclamation points
> as haphazardly as you do.

Therein lies the rub: What to do about a piece of punctuation that until the advent of texting and emails was used sparingly, or not at all? One that most people actually looked *down* on? Arguably, the return of the exclamation mark is one of the most epic comebacks in punctuation history—and it's a cautionary tale for those of us who don't intuitively keep up with the times.

Traditionally, exclamation points have embodied three basic meanings: urgency, excitement, and emphasis, qualities signaled offline by furrowing or raising our eyebrows, tapping our fingers, speaking rapidly, or, for the truly excitable, bouncing up and down on our heels.

Today, exclamation points, arranged throughout texts and emails, convey friendliness. They have become so obligatory in emails that you risk coming off as brusque or cold if you fail to use them. An exclamation point at the end of an email's opening sentence establishes a heartfelt sentiment that resonates through the rest of the message.

> @springrooove
>
> Adult email culture is ending every sentence with an exclamation point, then proofreading to see how many is socially acceptable to keep.

(Anonymous. Twitter Post. February, 20, 2019, 3:42 p.m. https://twitter.com/springrooove/status /1098337153648611329?lang=en)

At times, today's profligate use of exclamation marks is a way of attracting and maintaining a reader's attention. In essence, exclamation marks call out, "Yo! Listen! I'm talking to you!" Among digital natives, however, their use is far less tortured, or meaningful. They are an almost obligatory emblem of friendliness, more "I come in peace" than "There's a foot-long rat in our garage!" Women use a lot more exclamation points than men do, as the exclamation point serves as a text-based version of the nods, smiles, and laughs that typically suffuse female friendships. As my friend Karen once told me, "Sending an email without at least one exclamation point is one of the most terrifying and exhilarating things to do as a woman."

Of course, like everything, it can go too far. For example, it took two weeks for Bella to respond to Sheila's email about a new team effort. *Sorry for the delay!!!!* Bella finally wrote. Sheila, put out and feeling a little huffy, didn't even bother to respond. "Four exclamation marks with a 'sorry for the delay'? What a fake." From Sheila's perspective, too many exclamation points was a sign of phoniness, and false enthusiasm. But is it?

Although we generally interpret exclamation points as positive, most writers, editors, and traditional style manuals recommend using them as discerningly as ever (meaning we should try not to use them ever). If you do, and today we all do, you should use them judiciously, since in serious situations they can be interpreted as overly intense and even immature.

EXCLAMATION POINTS: A PRIMER!!!!

Understand—Really Understand—What an Exclamation Point Can Mean. In general, we use exclamation points when we want to say something extra loudly, or even extra nicely. "The exclamation point is the quickest and easiest way to kick things up a notch," writes Will Schwalbe, co-author of the book *Send: Why People Email So Badly and How to Do It Better.* An exclamation point adds velocity to your words while also serving as a sincerity marker, a stand-in for "I really, really mean what I just said!" This is especially true when three or four exclamation points are used in sequence: *Are you being sarcastic? No!!!!* Be aware, though, that exclamation marks can also mimic shouting when used with capitalization, or in a stressful context: *NO!!!!*

In my experience, using more than a single exclamation point can get tricky.

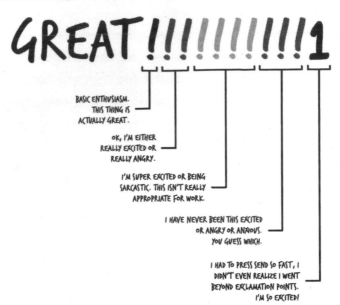

In short, with exclamation marks, it's safer to err on the side of minimalism.

The Pressures of Being a Woman!!!! Research shows that women feel compelled to use exclamation points to come across as friendly, warm, and approachable, whereas men are more likely to use them to signal urgency.[6] *Not* using softening exclamations can give rise to alarm. One female salesperson I know had a habit of responding to team members' emails with *OK*, leaving them to second-guess whether she agreed with them or was quietly seething at her desk. The simple awareness of this perception (and a change to *OK, great!*) infused her team dynamic with higher trust and camaraderie. We'll dive into this subject further in chapter 8, "Gender: *He Said, She Said, They Said.*"

TEARS, TONGUES, WINKS, AND FROWNS: AN EMOJI PRIMER

Beyond the simple smiley face, emojis provide texture and context to our sleek digital communications. What are the body language equivalents? Quite literally, our faces. In the real world, we supplement our expressions with hand and arm gestures and the tone of our voices (high-pitched and happy; gruff and angry; excited and enthusiastic). Emojis are nothing more than little faces designed to mimic the emotional range of our human ones.

In 2015, the Oxford English Dictionary revealed its word of the year: "face with tears of joy emoji," otherwise known to us as 😂. The decision was met with mixed reviews. Some said that declaring a goofy smiley face a "word" was an affront to the English language. Others welcomed it as the first step toward developing a universal language. I believe emojis are critical to enhancing workplace efficiency and cultivating a corporate culture of optimal clarity—and that even executives should use them to clarify their tone in office interactions.

Today, for even the most skilled communicators, emojis have become an essential shortcut. Not only do they appear in texts and group chat tools, but they also show up in PowerPoint slides, video meeting discussions, and emails. Using emojis, we can express ourselves faster, more vividly, and (literally) more colorfully. That said, we often create more confusion than we intended when we rely on emojis in lieu of actual words.

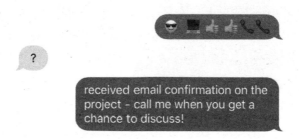

Humans send 6 billion emojis every day, with the average person sending 96 emojis over a normal 24-hour period.[7] Along with the Oxford English Dictionary, academia has taken notice.

"We're in a new phase of language development," says Susan Herring, professor of information science and linguistics at Indiana University. "More and more graphical representations, such as emojis, GIFs, stickers, and memes, are being incorporated into language used online."[8] In 2015, Chevy came out with an all-emoji press release.[9] An obvious gimmick, it nonetheless showcased how emojis can be used as a more or less universal language.

That said, emojis aren't going to become anyone's first language anytime soon. They're still closer to slang—best used to accentuate, not replace, actual words. And if clarity is our goal when we communicate, emojis may not be as universal as we believe.

First, it's extremely important to tailor emoji use to your audience. When is a smiley face just a smiley face, and when is it something else entirely, like a binding future commitment? When Israeli landlord Yaniv Dahan got a series of smiley texts from two prospective renters, he was convinced he had found the ideal tenants. They were both so positive and enthusiastic! Their texts radiated optimism and even joy—a smiley face, a bottle of champagne, a couple waltzing. After a few more of these emoji-heavy text exchanges, Dahan took his apartment off the market and waited for the couple to sign the lease. And waited. Eventually it became clear that Dahan's ideal tenants had waltzed away, probably with grins on their faces and a bottle of champagne under their arm. Dahan had been ghosted.

Unlike most people, Dahan refused to let it go. He appealed to the law by taking his emoji-loving tenants to small claims court, where they were fined $2,200. The judge put it very simply: the couple had "acted in bad faith" by using emojis to "lull" Dahan into a false sense of security. The judge added that "[t]hese symbols, which convey to the other side that everything is in order, were misleading, since at that time the defendants already had great doubts as to their desire to rent the apartment."[10]

In other words, sending the wrong emoji can cost you not just in miscommunication but also in cash. So be careful.

When deciding whether or not to use an emoji or wondering how to interpret them, consider the following: How comfortable are you using this symbol? A team I worked with once had four members collaborating on an international research project via group chat. Every time James shared a new insight, Ivy responded with a smiley face emoji. Every. Single. Time. These short symbolic responses made John doubt Ivy's intentions. Was she *really* that excited, or was she being sarcastic? For her part, Ivy was simply using a shorthand that was common with her Chinese friends and trying to show her support.

EMOJI WITH CARE

Don't believe the stereotypes. Emojis aren't just the province of "young people." Sure, kids and teens need to differentiate themselves as individuals, an impulse that historically gives rise to free-flowing and creative use of language. But older generations almost always end up adopting the vocabulary of younger generations, and emojis are no exception. "Super," "My bad," and "Awesome" sounded flat-footed or juvenile at first, but today they're used by all ages. Any work environment that puts up with "My bad" won't have problems with a senior executive using a smiley face.

Think before you emoji. Understand that depending on your gender, your culture, and your country of origin, the use of emojis will be received differently. One recent study showed that the overuse of emojis implied incompetence at work, and that younger women were more likely to be unfairly implicated.[11] In Western nations, the thumbs-up emoji signals agreement or approval, whereas in Nigeria, Afghanistan, Iraq, and Iran, it means "sit on it" and is considered vulgar, offensive, and not very nice in general. Along with dialects or regional accents, the use of emojis is generally taken to be a signifier of "geography, age, gender, and social class."[12] Some countries, for example, take the eggplant emoji for what it is, an odd-looking, blue-black vegetable. But in other countries, like the United States and Ireland, the eggplant emoji is considered a stand-in for the male penis. We'll decipher the world of emojis in more detail in chapter 9, "Generation: *Old School, New School.*"

GETTING DOTTY WITH PERIODS

Once upon a time, the period, along with the comma, was arguably the world's dullest piece of punctuation, used exclusively to end a sentence just like this. These days, more than any other punctuation mark, the period—and yes, we're talking about the same black dot—has evolved to the point where it now signals something else entirely, in this case cold, cruel fury, no different, really, from a furious facial expression.

Unlike any other piece of punctuation, the period has taken on an outsized, exaggerated, and often-unintended meaning

across digital communication. Imagine that someone texts you, *Can you watch my dog for the night?* Your reply, *Sure*, sounds a little iffy and on-the-fence. By contrast, *Sure!* conveys excitement and even eagerness (you love that dog!). Bringing up the rear is *Sure*. Sure. Yeah. Huh. Sure. What to make of that *Sure*. other than something like, "I could watch your dog, but I'll have to cancel my dinner plans, and you're being a pain to ask me, but I'll do it because I'm your friend though I'll resent you the entire time, and oh, also? You *owe* me."

In 2016, psychologist Danielle Gunraj ran a study testing how a group of research participants perceived one-sentence text messages ending in a period,[13] then contrasted her findings with how this same group perceived periods that showed up in handwritten notes. Gunraj found that in text messages, sentences ending with periods were more likely to be considered insincere. As far as the handwritten notes were concerned, the periods had no effect on perceived sincerity.

These same findings don't extend to emails. In email, periods can be used in the same way we use them offline without seeming angry or insincere.

My friend Aria is the CEO of the nonprofit DoSomething .org. Known for her upbeat, friendly personality, she uses emojis, exclamation points, and even the occasional GIF with her staff. Recently, in a hurried response on the team's Slack channel, Aria responded with a simple *ok*. No problem, right? Later that day, Aria's assistant told her that her colleagues had found Aria's *ok*. "absolutely chilling." Everyone assumed Aria was furious with them. Believing they understood her digital persona, with a single keystroke Aria had thrown them all for a loop. Bottom line: when a friend or colleague ends a text with a period, it's usually seen as aggressive, a cause for alarm.

DOT DOT DOT . . .

If a period is cause for hair-pulling, then an entire sequence of multiple periods, known formally as an "ellipsis," can cause even greater confusion. Do ellipses imply the person is asking a question? Making a statement? What exactly are those three dots supposed to mean? If you receive a message containing an ellipsis, are you supposed to infer magically what the sender is saying? Does the answer vary?

Unfortunately, the answer to all of these questions is *Yes.* Generally speaking, ellipses signify either the omission of information or the expectation that someone will follow up with a question or statement. As a word and piece of punctuation, *No.* shuts down a conversation fast, whereas *No . . .* leaves it hanging, and to be continued. In some cases, I've seen ellipses used as an instrument of hostility, an invitation to the offender to speculate about his faults and fix them (*Not sure if my email made it to you as I haven't heard back . . .*). Other times, they're used to convey humor or sarcasm (*Those glasses, tho . . .*). Still other times they signal, "Hang on," "Um," or "I don't know."

WHAT DOES THE ELLIPSIS MEAN?

The ellipsis is the most passive-aggressive punctuation mark, so use it cautiously. Hesitation, confusion, apathy, we'll talk about it later—they're all conveyed by those three dots. Ellipses suggest that something is going on but leave you wondering what that something is. To achieve optimal clarity, avoid them unless they effectively signal an unfinished thought.

Older people use them differently. Why do older generations use ellipses so much? It's already hard enough to determine the tone of texts, and then they hit you with *lol* . . . or *hello* . . . Especially for older generations who typically steer clear of exclamation marks, ellipses feel like a softer stop than a period since ellipses . . . drift away. But for digital natives (those born after 1985), ellipses can convey a whiff of sarcasm when read online.

SORRY, BUT *WHAT* WAS THE QUESTION?

As everybody knows, in nondigital writing, the question mark signals interrogation, interest, and even frustration, the equivalent of cocking our heads to one side and narrowing our eyes.

Historically, the question mark has always been called upon to navigate a certain tension. For example, Fernando is eating lunch at his desk when his boss sidles up to him. "What are you up to?" the boss asks. Fernando's first thought: *Is my boss genuinely interested in my life, or is this really a shady way of saying, "You don't seem to be doing anything"*? Maybe his boss meant, "You shouldn't be doing this," or even, "I have a task for you." Now, imagine the difficulty of understanding the subtext of a question mark communicated via email or chat, lacking, as it does, the accompanying facial expressions, vocal tone, or obvious body language of a person standing beside your desk.

What if the question is supplemented by three question marks instead of one? What about five question marks????? Multiple question marks convey urgency, impatience, and possibly panic. For example, if your friend texts you, *Are you at your desk?* you would think nothing of it. She's probably nearby and thinking about stopping by to say hello. But if the text read, *Are you at your desk???????* your stomach might drop.

In general, the more question marks there are in a message, the more intense the emotion likely is behind the question, especially among women. (We'll learn more about gender differences in chapter 8: "Gender: *He Said, She Said, They Said.*")

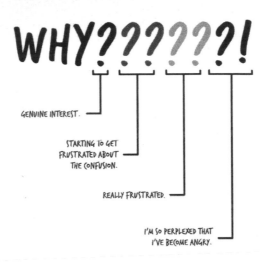

GOING REALLY, REALLY BIG: CAPITALIZATION

When we're feeling really excited about something or impatient for something to happen, how do we show it? If you're like most people, you might tap your fingers to signal "Let's *go!*" or raise your eyebrows, or clench your jaw, or even, I don't know, *holler.* Absent these expressions of physical body language, we are left with ALL CAPS.

We might interpret an email written in ALL CAPS as hostile, menacing, or indignant, but what if it was written by your grandma who only discovered email two months ago? Generally speaking, who you are *behind* the screen comes out in your writing *on* the screen. If two people enjoy high levels of familiarity and trust, in addition to a shared context, most likely they will interpret each other's messages with minimal misunderstandings—even if one receives a message in ALL CAPS. Consider the exchange below:

● **John**   10:02 AM   >
CAN YOU SEND THIS TO ME TODAY
This message could signal urgency or anger.

● **John**   10:02 AM   >
WHAT DOES THIS MEAN???
This message could signal interest or frustration.

● **John**   10:02 AM   >
WE NEED TO TALK
This could be an urgent meeting request or simply fast typing.

In order to decipher the capitalization in John's messages, we first need to understand John's relationship to the recipient. If John and the receiver are peers, then clearly John is a fairly entitled guy. If John and the receiver are close friends, then John is probably being casual and typing very fast. If John is your boss, well, shit, I'm sorry.

HOW TO USE ALL CAPS

- To avoid anxiety on the receiving end, try to limit the number of messages you send in all caps.
- If you type in caps at work, people are likely to assume you are YELLING, which is only good for comic relief.
- Prioritize using all caps only in urgent situations with your team.

Then there's "mixed punctuation." Riana receives a curt message from her boss, Theresa, saying, *Riana-Can you please NOT send those emails without approval from me?!?* Riana is confused by not one but *four* things: the hyphen, the italicized font, the capitalization, and the *?!?*

Why? Well, her boss seldom uses these flourishes, leaving Riana to interpret her email as disproportionately angry. Later, Riana finds out that Theresa was lashing out at anybody within email earshot after losing a big client—and that none of it had *anything* to do with Riana.

We've all dealt with people who get triggered and proceed to lash out indiscriminately at the closest target using whatever tools are at their disposal—a phone, a pen, a flower vase, or, in this case, punctuation. All that emotion, encapsulated by hyphens, italics, question marks, and exclamations!

TIMING—THE NEW MEASURE OF RESPECT

In 2017, Paige Lee Jones (@paigeleejones) aired her biggest pet peeves on Twitter: "Got an 'out of office for the holidays' email after responding to a requested URGENT email within 4 minutes."[14]

When we speak face-to-face or on the phone, it takes us an average of just 200 milliseconds (that's 0.2 seconds) to respond to another person.[15] It's also clear to most of us when a conversation is over—we've walked away or hung up the phone.

But face-to-face and phone exchanges require both parties to be available at the same time. This is less possible today, with most of us scrambling through our days and some of us collaborating with colleagues across multiple time zones. This, in fact, is a key benefit of digital communication—we don't *have* to sync up at the same time or place to engage in a real-time dialogue. It takes 90 *minutes* for the average person to reply to an email, and 90 *seconds* for the average person to respond to a text message.[16] Digital communication allows us to interact with others at our convenience, but that also

means it can be S-L-O-W. If we are being honest with ourselves, most of us are uncomfortable with pauses and silences.
What's with all this quiet? Is everything okay? Our brains come
up with one explanation after another to explain the absence
of an immediate response, especially in situations where trust
is low and power dynamics are out of balance.

Digital conversations are often *asynchronous*, meaning
that you and I aren't necessarily having a conversation in "real
time." For example, I could send you an email as you're running on the treadmill at the gym. I may have just started our
conversation, but you aren't likely to engage in it for another
hour, two hours, three hours, or more. Asynchronous conversations give us more control over when and how we respond,
but if you're the one who's waiting to hear back, the gaps in
response time can produce anxiety. Responding to your employee's urgent text for help five hours later has the potential
to leave your employee feeling angry and alone. As for that
little bubble with three dots in iMessage? Yes, it's handy for
telling you when someone is typing, but when it just sits there
like a pulsing heart (or is that *your* pulsing heart?), every milli-
second can last an eternity. Then, when it suddenly disappears,
you're left wondering whether you are being ignored or forgotten, or whether someone better came along.

In a digitally reliant world, the slightest pause between
messages takes on an almost operatic meaning. The thing is,
most of the time a non-answer means nothing at all; the other
person got tied up, was doing something else, didn't notice
she'd gotten a text, had her volume turned off, or forgot where
she put her phone.

One night my phone died on me when I was running late
to a dinner with friends. My friends, who were already at the
restaurant, grew increasingly worried—to the point where one

called my husband, who in turn tried reaching me multiple times, with each call of course going to voicemail. My husband works for a bank, and in his world everything demands a quick response (his cell phone never falls below the 50 percent charge level). He was so worried and upset that he left his *own* dinner and raced across town to meet up with *my* friends so that together they could try to hunt me down.

Can you imagine this scenario taking place just ten years ago? It didn't, because a decade ago if you didn't hear back from someone after an hour, you waited another hour—and, well, the world went on.

Contrast that scenario to today, when digital silence has taken on new, potentially threatening meanings. In the workplace, more than making people feel worried, silence often makes our colleagues and co-workers feel snubbed—especially if a "read receipt" shows that the message has been read and the other person has not responded. As a friend of mine once put it, "I never know if they've actually read it. If you have, then why aren't you answering? Are you upset with me? Are you ignoring me?" When asked to consider the possibility that the receiver is busy or needs more time to respond in a thoughtful way, her response was, "Yeah, I guess. I just know a lot of people that don't respond to let you know they're mad at you."

Quite possibly, the people you believe have it out for you aren't mad at all. They're usually not even thinking about you! Consider the possibility that the person on the other end of your communication is simply overwhelmed. As my client Sarah put it, "Sometimes I don't answer because I don't have time to give the response I think is deserved, so I put it off until later. Then I forget, and [the other person thinks] that I didn't care enough to respond, when, in fact, I cared too much."

Adds Adam Boettiger, a well-known digital marketing consultant, quoted in the *New York Times*, "We've seen an increase in the nonresponse rather than just politely declining. You delete it and hope it goes away, just like if someone comes to your door and you pretend you're not home."[17]

WHAT IS AN ACCEPTABLE RESPONSE TIME?

- It's generally considered acceptable etiquette to wait up to 24 hours before responding to an email.
- For texts and instant messages: respond quickly during business hours, or risk being considered rude and offending the other person.
- If you receive a message outside reasonable work hours, feel free to ignore it until those hours come around again. If this happens rarely, consider responding with a quick message alerting the sender to the fact that you won't be responding until later. It's better to reply with a quick *Got it! I'll get back to you by Tuesday* than to leave the recipient waiting until you provide a full response.

Each medium—texts, emails, phone and video calls, and all the rest—comes with its own built-in timer. Emails are faster than calls. Texts are faster than emails. Despite constantly having our phones with us, there *are* ideal times to place a call. When calls aren't scheduled in advance, place a call at the :20 or :50 minute mark of an hour, when others are usually finished with other calls planned at the hour or :30 minute mark. Weekdays during normal work hours, especially mornings, are the best times to send an email that will get a reply and schedule a video call. On weekends and afternoons, prepare to receive shorter replies.

It's perfectly reasonable to set your own boundaries and communication norms.

We can help ease anxieties around these timing expectations with simple, clear communication. When you send an email at an inopportune time, refer to it with a simple *No need to respond until the morning.* If you are responding late, consider directly acknowledging the timing gap: *Thanks so much for your kind note last month! Things have been crazy here ever since, which is why I'm so late in answering your email. I apologize!* For important work emails, be heartfelt: *I am sincerely sorry for letting the ball drop on this one. In the future, I'll double-check that I've sent my messages to you so it doesn't happen again.*

TO, CC, BCC, AND REPLY ALL—
THE NEW MEASURE OF INCLUSION

Think of an email as a sporting event. You and whoever else is in the "To" box are the athletes. If you don't cc or bcc anyone, you're just practicing, rallying before a match, or throwing the ball around with a friend. When you add observers to the "Cc" line, suddenly other people begin to fill the stands. Add more people to the "Bcc" line, and you're now swelling the VIP box seats with scouts, coaches, and recruiters. From here, the stakes go up. If you choose to reply only to the other athlete, you're having a private conversation no one else can hear, whereas "Reply All" is equivalent to a booming voice coming in the overhead speakers that the entire stadium can hear.

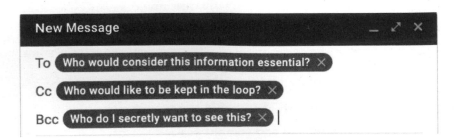

Cc and Bcc use is inherently tricky. One of my clients, Janine, once put it this way: "I often want to share an email with my boss for her reference. But Bcc feels like I'm inviting her to spy on my conversation. And cc makes other people feel like I'm trying to show off or get credit. I prefer to send the original email and then forward it on to my boss. It makes it look like I sent it normally and then later felt the need to loop in my manager for visibility." Unfortunately, Janine's fear of judgment and her interpretation of Cc and Bcc ended up creating even more work for her.

Some of my other clients have a primal (and reasonable) fear of Reply All: "I prefer transparency," Steve explains. "My job requires me to send out group emails to get feedback. Yet no matter what I do, there are always a few people who want to Reply All—even when I start the email with a bolded, all caps statement saying, **'DO NOT REPLY ALL.'** So now I have to Bcc to protect everyone's inbox from being flooded with unsolicited Reply All feedback."

In one organization I worked with, I was reminded that clicking the wrong box can lead to grave consequences. For almost seven years, Corinne had been working at this firm, where she was regularly preyed upon by an especially cruel colleague whom we'll call Melissa. Corinne was hardly the exception; teams across the entire organization were aware of Melissa's reputation for being vindictive. But Melissa wasn't going anywhere, at least not for a few years until she retired. Working with Melissa took its toll, but Corinne loved her job and decided to stick it out.

One Friday night, while working late, Corinne received some great news. Management had finally caught on and was planning to fire Melissa. Corinne did her best to disguise her excitement, and at the same time couldn't wait for everyone's reaction when they returned to the office on Monday morning.

On Sunday night, Corinne reviewed her calendar and email on her phone to prepare for the upcoming week. A message came through with the subject *Exciting Update!* in reference to a new project the firm had taken on. Still giddy from the news on Friday, Corinne impulsively forwarded the email to her closest friends at work, adding, *Ha! I thought the exciting update was going to be about Melissa getting fired!!*

Two seconds later, Corinne's email landed in her own inbox. Quite rightly she panicked. With growing dread, she scanned the address line and below it her remark, *Ha! I thought the exciting update was going to be about Melissa getting fired!!* She had messed up. She had pressed *Reply All*, "all," in this case, meaning more than 300 people at their 500-person organization.

Frantically, Corinne tried to pull the email back by calling a friend in IT to get it recalled—no luck. Next she tried calling and texting her boss—no answer. She spent a painful, sleepless night, arriving early the next morning to work only to be told to go directly to HR, where she was fired, just like that.

It didn't take Corinne long to find a new job, but her experience with Reply All left a lasting scar. These days, she says, "I triple-check every email I send, and I definitely don't use my personal phone for work!"

Reply Alls, Ccs, and Bccs are necessary in most workplaces, but ask yourself who really needs to be included. This involves discernment, because some people insist on being a part of everything. Reply All should be limited to high-priority information you want to share with the entire team:

meetings, announcements, agendas, and enterprise-wide information. Always be conscious of the power dynamics and trust levels among your recipients—and avoid jumping to conclusions when you receive a message that catches you off guard.

YOUR DIGITAL PERSONA— THE NEW MEASURE OF IDENTITY

Before I hire anyone, whether it's a babysitter or a marketing consultant, I google them. Most employers do this, but more and more I hear stories of peers doing it to one another, and parents doing it to their kids' friends. My neighbor once confessed she even googled our doorman. In a digital world, who we are online is likely the first impression we show the world and, just like in real life, first impressions matter.

Let's break down the main components of our digital personas:

Your Name. Sometimes it really *is* all in a name— especially if you work across teams you've never met face-to-face. If all I know is your name from an email, calendar invite, or Slack channel, what conclusions might I draw? Let's say your name is Maxine. Do you go by Maxi (which suggests you may be more casual) or by the full Maxine (which points to formality at work)? What if you go by Max (signaling you could be female, male, or gender nonbinary)? The recipients of your messages create an image of you based on your name, so choose wisely. Across social media, use just your name, no funky middle names, no movie character names, etc. Just be you.

Your Email Address. Do you have a Yahoo!, Hotmail, or

Gmail account? Are there numbers after your name, signifying a potential outdated email address? Do you use a personal email or a company-branded one? Your work address may signal your power level, and might also explain why your emails sound so formal. In addition, giving out your personal email address may imply a desire to stay in touch for the long term outside of work.

Your Profile Picture. Is there a photo that comes up alongside your email address on Outlook or Gmail? Do you have a photo set on your Zoom or Webex profile? A picture of a sunset tells me a lot about the sunset, and almost nothing about you. It's good to put a face to the messages people see by adding a clear, professional picture. The quality of the image matters, too. A low-resolution photo gives off a negative impression, whereas a high-quality photo signals that you are strategic about your brand.

Your Search Results. After plugging in your name, what are the first three websites that come up? Do you have your own branded website? Are you highlighted on your company website? Were you quoted in your local newspaper on a politically charged topic? All that gives me a sense of who you are. Ideally what people find should tell them that you're professional and trustworthy, the opposite of someone with, say, easily searchable mug shots. Make sure your LinkedIn is updated, so that people can easily find things about you in your professional life.

· · · · · ·

Now that we've discussed the many forms that digital body language can take, part two of the book will show you how each of these new signals affects teams holistically—how we can use digital body language to show appreciation (Value Visibly), to find alignment (Communicate Carefully), to re-define teamwork for a digital era (Collaborate Confidently), and, lastly, to put those three pillars together to form teams characterized by psychological safety (Trust Totally).

The Four Laws of Digital Body Language

THE FOUR LAWS OF DIGITAL BODY LANGUAGE

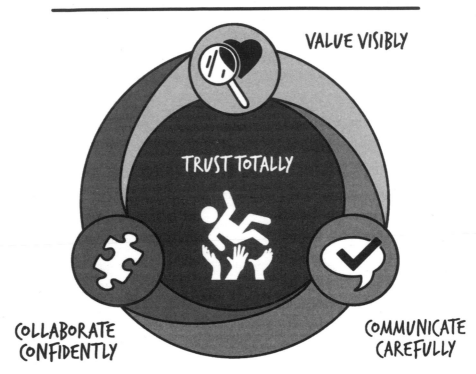

VALUE VISIBLY

TRUST TOTALLY

COLLABORATE
CONFIDENTLY

COMMUNICATE
CAREFULLY

·

CHAPTER 4

Value Visibly

Stop Disrespecting Me!

Remember how it felt when someone looked you in the eye, gave you a firm handshake, and said with feeling, "Thank you so much"? In the digital workplace, we **Value Visibly** by being attentively aware of other people and clearly communicating "I hear you" and "I understand you" by using the new cues and signals of our digital body language. Valuing Visibly means being more sensitive to other people's time and needs, reading digital communications with care and attention, and respecting other people—without being in a rush about it.

I'll call him Jim. I was in New York City, and he was in Dallas. I had just started my own consulting business, and Jim was just beginning his post-college career. During our video interview, he responded quickly and intelligently to my questions. He seemed interested in the work I was doing on collaboration, and he was a good communicator too. Eager to get started, I hired him on the spot as my new marketing strategist.

I chose well—Jim was a fantastic employee. I loved what a self-starter he was, and how little oversight he needed. As I rushed forward in my business, he always matched my pace, finishing whatever administrative tasks I texted or emailed him without requiring much guidance. Whenever I sent him a work request, Jim's usual email quick response—*Sounds good*—gave me confidence that he had everything under control. (And my typical *Thx* response made me feel like he knew I appreciated it.)

Fast-forward to our phone check-in six weeks later:

> Me: So, how do you think it's going? Because I think it's going really well!
>
> Jim: No, it's not.
>
> Me: No, it's not? Wait, what?
>
> Jim: No, it's not going well and I'm thinking of quitting. Today.
>
> Me: Today? Wait, sorry, *what*?
>
> Jim: Look, I have my master's degree, and I don't want to do only admin work. I thought I would be handed a lot more marketing. Like we discussed in our original interview? And we barely talk about what you're working on and the larger picture of the business.

It turns out that as I was busy powering forward, for the most part relying only on weekly phone check-ins, Jim was sitting there in Dallas, stewing, uncertain about how he was doing in his job and—even more importantly for a young professional— not knowing exactly *why* he was doing it. His *Sounds good* emails didn't signal, "I will take this on happily." They were code for "I will do this begrudgingly, but I really want to talk about my learning goals too." My *Thx* emails, intended to communicate "I really appreciate your hard work," were interpreted by Jim as dismissive. I thought I was being explicit, but in the end, Jim felt both underappreciated and disrespected.

Was I an inexperienced leader at the time? Absolutely. If Jim and I had shared an office, would I have treated him in the same way? There's no way. Looking back, I realized I'd showed Jim a lot of disrespect. I was always 8 to 10 minutes late to our weekly calls, wasting his time and angering him with every additional *sry running late* email. Sometimes I ended our phone conversations to take another call without first explaining why I needed to answer the second one right away. By the time I turned my attention back to him, both of us had lost our train of thought, forcing us to waste even more time picking up the flow of our discussion. Lastly, and most embarrassingly, I would send incomplete emails, responding to *some* of Jim's questions but not all, since I was otherwise busy prioritizing more urgent tasks.

Jim was right to advocate for himself, and to be honest. He also offered me a chance for redemption. Our resulting conversation, as uncomfortable as it was, reminded me how important it is to Value Visibly, to consciously, outwardly show respect toward people in our professional and personal lives.

Over half of all employees report they don't receive the respect they need or want from their leaders.[1] That sounds like a whole lot of ungrateful leaders! But could there be another explanation? What if those leaders are expressing respect in ways that some employees don't recognize? As the signs of respect have changed, so too have the skills we need to use to make our colleagues feel valued.

Traditional respect has always been based on signals we share in person. Each personal interaction generates positive signals that our brains understand unconsciously after hundreds of thousands of years of evolutionary training. But today, as I note throughout this book, many of our interactions lack visible cues of meaning and understanding.

....

When relationships are mediated by screens, how do we make the invisible visible?

....

After talking to Jim, I realized I was making a mistake that is also one of the greatest killers of engagement today: I assumed that if I didn't hear anything from Jim, everything was okay. ("No news is good news," as the saying goes, even though that's no way to run a business.) With up to 60 percent of teamwork today conducted digitally and via the written word, we can no longer rely on assumptions to gauge feelings of mutual respect.[2] My lack of face-to-face interaction with Jim meant I was missing out on a lot of important information. I had to move from relying on a single conversation per project to multiple touchpoints and from *unstated* appreciation to *stated* recognition.

After our conversation, I made sure to end previous meetings on time to avoid causing delays. Jim and I scheduled weekly video check-ins to review work and ensure he felt valued and supported. The video format gave me an opportunity to read Jim's body cues so I could observe firsthand what made him uncomfortable and see his body language when he seemed unable to express himself in words or needed more time to reflect on something. Well, "sitting" with someone without rushing back and forth via email solved our communication challenges pretty quickly. We also had the opportunity to discuss Jim's professional learning goals, ultimately creating a project that he would complete in conjunction with his other responsibilities.

In general, I got better (*a lot* better) at checking in with Jim, giving him feedback on his contributions, and expressing appreciation for his hard work. End result? Eventually, Jim and I worked together for years. And the lessons I learned still guide me.

Today I leverage digital tools like video calls and weekly or biweekly email "check-ins" to ensure that my team feels Valued Visibly. For each person, I choose a communication medium based on his or her personality style (e.g., my college interns communicate best on Slack and enjoy emailed Amazon gift cards at the end of their tenures, while my executive team sticks to emails and appreciates personalized notes) while making sure I express encouragement or gratitude as often as possible. I don't leave messages unanswered, and when meeting with my team, I don't allow myself any other digital distractions.

> **Valuing Visibly means not assuming people are "okay." Instead, it means being proactive about explicitly showing you understand their desires and value their participation.**

As my experience with Jim shows, respect enables leaders to challenge a situation, not a person, and creates an environment where team members feel valued to engage in healthy, and even heated, conversations. Respect facilitates innovation and creativity by allowing leaders to tap into the power of diverse thinking and multiple perspectives. On the other hand, disrespect (whether intended or not) is the silent killer of collaboration, initiative, and job satisfaction.

STOP DISRESPECTING ME!

In the middle of doing research for this book, I got a frantic phone call from a client. Their head of human resources needed a favor. *Fast*. One of the company's leaders was managing a high-profile project and was having trouble keeping teams up and running. It was a startup of sorts, so people were expected to work long hours based on the expectation of future equity. But the teams were *not* collaborating, and communication was at a standstill. Everyone felt it too, which was affecting morale as well as speed to market and customer deadlines. Could I help?

Of course! I said, adding that even though I was on a research sabbatical, I would be happy to speak with the executive and create an action plan. The executive and I had a long, informative discussion during which we talked through some solutions. At the end of the conversation, he seemed eager to get moving and asked me to send him a proposal within a week, and to plan to start working together within three weeks. Absolutely, I said. Working late for the next few nights, I delivered the proposal on deadline.

Then—nothing. I never heard from him again. I later met other consultants who'd had the same experience of getting caught up in an enthusiastic conversation with this same person, then getting ghosted after delivering the promised work. I could only guess that he treated all his teams the same way— with a total lack of respect for their time and expertise.

Value Visibly is easy to preach but a lot harder to embed in the modern workplace. There is no shortage of articles talking about establishing respect by building a code of ethics or remembering to greet people in the elevator—but how does that translate to emails, IMs, and conference calls? The inherently distant nature of these communication channels makes disrespectful behaviors easy—even in live meetings.

I'll never forget a meeting I once had with Michelle, a senior executive at a large company. It took five emails, two follow-ups, and one confirmation call with her assistant to find a time that fit her calendar. I showed up at the appointed time. Nearly ten minutes later, Michelle walked into the room, greeted me, and immediately said, "You picked the worst time for this meeting. I have a big presentation I have to give later today." I offered to reschedule, but instead Michelle asked her colleague to come in and take her place in the meeting while she stayed there prepping for her next meeting on her phone.

The whole situation was bizarre. I felt more disrespected by Michelle sitting in the room on her phone than I would have if she had just excused herself and allowed her colleague to take the meeting. I couldn't help but be reminded of Jim, the great employee I had once almost lost. That feeling of being undervalued and disrespected stays with you. Why would I ever recommend Michelle to anyone in my own network? Who knows how many great opportunities she's lost over the years as a result of treating others so poorly?

COMMON PET PEEVES TO AVOID

- **Being in a rush.** Sending a message without proofreading. Trying to speed through a conference call in order to get to the next one. Claiming you're "too busy" to check in with your teams.
- **Not respecting others' time.** Double-booking meetings. Prioritizing your own schedule over other people's during the scheduling process. Letting conference calls run overlong. Sending "urgent" emails that are in no way urgent. Allowing useless recurring meetings to remain on the calendar.

- **Forgetting to show gratitude.** Getting into the habit of written-only communication without including phone or video check-ins where teams can actually hear "Thank you." Sending vague emails. Not crediting everyone on the team when sending in a deliverable.
- **Multitasking during face-to-face and video meetings.** "Just answering a quick text..." routinely halfway through meetings. Responding to emails and IMs on your laptop. Looking down at your phone when others are trying to make eye contact with you. Not putting your notifications on silent or vibrate mode during important discussions.

Moreover, in the digital world, it sometimes seems like we have too many options—and an equal number of opportunities for potential disaster. When should we email, and when is it better to text? When is a phone call expected? How long should we wait before replying to a message? What's the right time frame for digital thank-yous or apologies? Too soon and we risk seeming slipshod or insincere; too late and we risk coming across as unfeeling. Do digital thank-yous and apologies carry as much weight and import as those delivered in person or by phone?

Today, it's no longer safe to assume that someone "gets what we mean." That also includes whether they feel we are valuing them visibly or not.

VALUE VISIBLY: THE PRINCIPLES

Considering our probably permanent transition to digital communications, more remote work, flatter teams, and a more accelerated pace of change, the new principles of Valuing Visibly have never been more critical.

Reading Carefully Is the New Listening

By now you probably recognize that, whereas we once talked and shared information across a table or phone line, we now converse in written form. Instead of listening as others share their ideas, we read what they have to say in an email or on another digital medium. The problem, according to research done by linguist Naomi Baron, is that we comprehend less when reading on a screen than we do when reading print.[3] We devote less time to reading an onscreen passage, are more inclined to multitask, and tend to skim and search instead of reading slowly and carefully.[4]

For example, here's a recent email exchange I had with a client:

> do you want to speak Wednesday or Thursday?

> yes

It left me speechless. I'm *still* speechless!

● ●

Always reference details in your communications.
It shows that you put in the time to really read
through the message, think about the issues, and care
about the work the other person did.

● ●

One big reason we read so poorly online is that typically we're moving at lightning speed. Instead of taking the time to go carefully through messages, we race through them toward an indeterminate finish line (one that resets every morning).

Our need for speed leads to exchanges like the one above—the digital equivalent of talking over each other.

But are we *really* as busy as we think we are? According to Baron, we're just . . . not. A lot of our speed, and our anxiety around speed, is artificial, which ends up costing us accuracy, clarity, and respect. But even if you really *are* too busy to get back to people immediately, there are ways to show you aren't blowing them off. You can show respect, for example, by sending a quick note (e.g., *Got it!*) to let them know you got their text or email and are on it. You could give a ballpark estimate as to when you'll be able to respond at greater length. Ultimately, the goal is to show that you've really read other people's messages by addressing all their relevant points and answering any and all questions. If that's not possible, let your colleagues know you'll get back to them with more answers when the time is right. That way they know you're not ignoring the other items.

HOW CAN I SHOW ACTIVE LISTENING IN A DIGITAL CONVERSATION?

- Prioritizing quick response times, even if only to say you'll respond later.
- Answering all the questions and comments in the message, not just one or two.
- Asking *Can I call you?* or setting up a face-to-face meeting for more complex issues.
- Not interrupting—and stopping *others* from interrupting.
- Using verbal cues, such as "Go ahead" or "I'm listening," to encourage others to share their thoughts during video calls.
- Not using the mute button as a license to multitask.
- Asking clarifying questions.
- Taking notes or ensuring notes are sent out after the call.
- Giving teammates time to share ideas in a virtual chat box during a video call.

Writing Clearly Is the New Empathy

Writing well and, above all, *consciously* is a critical mark of respect. Period. (And no, I'm not angry with you.)

The CMO of a pharmaceutical company was communicating with her team about preparing a presentation for a board meeting. She shared a quick idea over email—*Do you think we should add more research on oncology to the presentation?* In her own mind, she was convinced she'd said, *Let's add an extra two bullet points on this slide*—but her brain was playing tricks on her. Two weeks later, her team had spent 30 hours or so preparing 40 slides on oncology research. The CMO had no idea the deck was coming and had frankly forgotten about the two bullet points she thought she'd proposed. But her team had gotten used to responding in full to her requests, and they seldom asked questions. Which made them feel even more devalued when their 40 slides turned into two bullet points on a slide.

Bottom line: if you're the boss, be mindful of writing "think-alouds," and separate them from true marching orders. If you're on the receiving end, don't be afraid to ask clarifying questions up front. A clarifying question is less embarrassing and time-consuming than a poor work product down the line.

· ·

When writing, do the little things. Check your tone, and think about how your message may be perceived, especially based on your rank.

· ·

A lot of the time, a misinterpreted email is the result of a dropped word or misleading punctuation mark. The solution is simple: proofread your emails! Proofreading is both a habit and a skill: making it a point of pride to send clean, unambiguous copy will help people take what you write more seriously.

....
A phone call is worth a thousand emails.
....

A German client once told me, "I was having a never-ending email conversation back and forth with a French and an Indian colleague who were having a circular written dialogue, covering the same ground over and over without understanding each other. I got them both to hop on the phone with me, asked a few questions a few different ways, and we got to the bottom of the issue. Sometimes, I think we are all guessing what the other parties mean on an email chain when we actually have no idea."

A good phone conversation is fast becoming an obsolete art. Which is too bad, since a call can save lots of time while simultaneously generating goodwill. (C'mon, we can't explain *everything* digitally!)

If you just received a vague or confusing text or email, don't be afraid to request a phone conversation or, if possible, a video or in-person meeting. If it's a sensitive dialogue, requesting a quick call shows you're being thoughtful. Instead of making you look indecisive, waiting a few beats before responding to questions shows the other person that you are listening and taking your work seriously.

With so many written platforms at our disposal, we can also get caught up in asking too many questions in email or group chats. Phone, video, or live meetings safeguard us from filling our inboxes with one tiny question after the next, instead requiring us to formulate the *right* questions.

At the beginning of any project, it's more helpful to ask open-ended questions than nuanced ones. "It helps me see whether the other person understood what I said," a leader told me once. Requests like "Tell me what success looks like for you," or "Help me understand what the best next steps are" shut down a slew of frenzied email chains, ensuring that everyone on the team is clear about the project goals and their individual roles.

VALUE VISIBLY IN ACTION

Practice Radical Recognition

The absence of respect can turn small details into big deals. Let me explain.

I won't ever forget a 30-minute phone call I once had with four colleagues in which the host waited until approximately the twenty-sixth minute to ask, "Does anyone on the line have any thoughts?" Up until then, instead of taking advantage of the four subject-matter experts he had on the call, he'd spent most of the meeting lecturing us! Not only did he come across as rude and self-centered (okay, he *was* rude and self-centered), but by not allowing anyone else to say anything, he was short-changing *himself.*

Whenever I host digital meetings, I usually ask remote attendees to lead parts of the agenda. They feel valued, and what's more, everyone can get to know one another's names, faces, and presentation styles. Typically I schedule a discussion based on pre-reads sent one or two days before the meeting. During my live webcasted workshops (where some attendees are in the room with me and others are watching online), I start my Q&A by asking virtual participants to share their questions first, making it a point to remind the people in the room that they are not the only ones participating in the workshop.

Anyone can create new norms and rituals to help ensure a company's culture puts a premium on recognition and respect. For example:

Scott Gerber, CEO of the Young Entrepreneur Council, sends video messages to convey his gratitude.[5]

A senior leader in China named Xu holds monthly 60-minute video conference calls with staffers in all branches to update them on business performance. Teams also use these calls to share their success stories. Most begin with a quick self-introduction by new attendees and a "celebration" of birthdays during that month. After six months, Xu is reaping the benefits. "People feel more engaged, more a part of the mission," he says, "because they know just how well they are doing at every level."

Aria Finger, CEO of the nonprofit DoSomething.org, rewards employees in unique ways. Among the handful of simple yet memorable actions she has implemented, she awards everyone who has been at DoSomething.org for three months "a personal emoji on Slack," and the company celebrates standout team players at an "awards" ceremony.[6] DoSomething.org has been named one of the Best Places to Work by *Crain's* and has a high retention rate among its nonprofit peers.[7]

Another executive I know runs an organization of over 1,000 employees. He calls every employee on their birthday, congratulates them, and thanks them for their hard work. As daunting as it may sound, the impact has been overwhelmingly positive.

Participants in a study published in the *Journal of Personality and Social Psychology* received one of two emails requesting help writing a cover letter. Half of them received an email with a line including the words *Thank you so much!* and the remaining half got the same email minus the show of gratitude. The study found that the recipients who received the *Thank you* email were more than twice as likely to offer assistance.[8]

These are all good examples of how we can show recognition in our digital body language inside our workplaces. It probably goes without saying that expressions of gratitude and respect don't have to be fancy or formal, nor do they require a lot of time. Typing just four extra words—*thank you so much*—can yield amazing results.

Acknowledge Individual Differences

One of my clients, Lisa, a technology executive, once shared with me the challenges she has in meeting the needs of both the introverts and extroverts on her team. "It's hard enough to manage the differences between introverts and extroverts with regular face time with my team," she told me. "Now I find my introverts won't jump in on phone calls or in rapid email exchanges because the louder voices still monopolize the conversation." Lisa also found that her whole team was less likely to share difficult news with her on team calls because they feared it would sound disrespectful, as though they were trying to throw others "under the bus."

To address this, Lisa has created a process following every monthly strategy call. She asks every team member to email her directly by the end of the week and answer two questions: "What's the bad news I don't want to hear?" and "What might we have missed in our last discussion?" She does this for a couple of reasons.

First, asking for bad news creates a regular space to speak up about challenges in the business. Second, Lisa's introverts require more time to process ideas, and they are more likely to speak up in an email or one-on-one conversation. By giving them the space to think through the questions, Lisa gets excellent insights she wouldn't have gotten in the meeting while reducing overall cultural groupthink. Lisa is also aware of the different ways team members engage in conversation and goes out of her way to meet them where they are comfortable— during a one-on-one post-meeting call, perhaps, or at a small-group lunch. Bottom line: *everyone* feels more respected.

HOW TO CONNECT WITH INTROVERTS

- Schedule down time between long meetings.
- Practice waiting five seconds before jumping in to speak.
- Send questions a few days before a meeting so they have time to process and prepare.
- Encourage them to email or message you with their thoughts after a meeting.
- Create a time limit so that louder voices don't monopolize the conversation.
- Stop interrupting. Use tools like a chat bar or hand-raising feature to designate who has the floor to speak, and choose a moderator to ensure that it's upheld.

HOW TO CONNECT WITH EXTROVERTS

- Set up regular face-to-face or video meetings so they can talk things through with you.
- Use breakout groups so they get a chance to talk out their ideas before reassembling as a larger group.
- Maintain watercooler spaces in your office or online so that they can recharge with social interaction between blocks of work.

Then there are *unusual* situations. Sue was the head of licensing at a publicly traded fashion house. Every quarter she met with Doug, the CFO, to review her team's budget. Four times a year, her staff worked long hours, planning and documenting every aspect of its complicated budget. While Sue and her team had a great deal of mutual respect, the same issue always came up at budget time.

It was this: Doug preferred to discuss the budget alone with Sue. Needless to say, this caused her team to feel excluded from the project they had all worked so hard to accomplish. Even more demotivating, there was never any recognition of the team's work or any explanation of subsequent budget changes. When Sue finally caught on to the problem, she initiated a couple of simple processes. First, she made sure that everyone's name went into the final presentation, which clarified who had created what. Following each and every quarterly budget meeting with Doug, she also scheduled an immediate one-hour meeting with her team, during which she reviewed Doug's comments and feedback. She also wrote the CFO a follow-up email that acknowledged her team's individual efforts, citing the contribution each person had made to the end product. Of course, she cc'd her team as well so they could witness her singing their praises.

In our increasingly competitive workplaces—where both the pace and the technology make it easy to lose touch with the human connection—this kind of valuing process makes a difference.

BECOME A MEETING NINJA

Valuing visibly requires you to "watch the clock"—and I mean that literally. This may seem overly tactical to some people, but I find that when you don't respect others' time in phone, video, or face-to-face meetings, it sends the implicit message that you don't value them at all.

Take Jonathan, who was invited to join a team conference

call the night before it occurred. Being added late made him feel overlooked before the discussion even began. To make matters worse, nobody told him what the meeting was even *for*. As the meeting began, he soon realized that most of the others had no idea why they were there either, who else was coming, how long the meeting was going to last, or why what was being discussed couldn't have been communicated in an email.

Five minutes into the discussion, Jonathan stopped the host. "Excuse my interruption," he said, "but before we really dive in, can you share what success looks like at the end of this meeting and what the agenda will be? And could we do ten-second introductions of who's on the call?" Immediately, a clear purpose was established. Everyone had clarity on the value that they were expected to provide to the meeting and what they should expect going forward.

How can leaders create meetings that value and respect their teams?

Design meetings with a clear agenda and plan to offer clear action steps at the end. This shows respect for your colleagues' time, while also communicating accountability. At the beginning of a meeting, say, "What success looks like for this meeting is XYZ . . ." At the end, recap if you've achieved that success, or list what's missing.

At the start of every meeting or phone call, set aside five minutes to make introductions. Ask everyone to share one personal or professional update. This allows for increased vulnerability, familiarity, and trust and helps everyone involved understand where their colleagues are coming from. Twenty-four hours before the meeting, distribute an agenda that encourages different members to lead a section of the meeting. Periodically ask for input instead of waiting until the end. Let everyone pitch in. If you're on a phone call, ban the mute button in order to minimize awkward pauses and multitasking.

Know when to exclude others from meetings in deference to their time. For example, one Fortune 500 chief digital officer regularly removes senior leaders from recurring meeting invitations when their input is no longer needed.

Put simply, we all value our time, and showing our respect for it has an outsized impact on people's happiness and overall commitment to work.

HOW TO MAKE MEETINGS MORE VALUABLE TO EVERYONE

- Make sure everyone can answer the question "Why am I in this meeting or conversation?"
- Schedule meetings only for the minimal time required. Consider Parkinson's Law: Work will extend to the time allotted for it.
- Start and end meetings on time.
- Send out a clear agenda or define the desired outcome before a meeting.
- Host a weekly one-hour "virtual office hours" session to handle smaller issues that don't warrant group meeting time.
- Audit recurring meetings and eliminate the ones that don't add value.
- Do not create a meeting with more than eight attendees unless it is a broader team strategy session, town hall, or division-wide update.
- If you have invited a senior leader to a meeting, clearly state whether their attendance is optional and whether a proxy is requested if the senior leader can't attend.

BAN MULTITASKING

It's 4:15 p.m. on Wednesday, and I'm responding to my emails on one tab, doing some early Christmas shopping on another, and choosing a restaurant for dinner on my phone. Just as I'm closing in on the perfect Christmas gift, a voice shocks me back into reality . . .

"Erica, what do you think? . . . Erica. Erica? Erica!"

Right. I'm on a *conference call.* "Sorry about that, I was on mute," I say, even though I wasn't on mute, I was just focusing on other things. What was everyone just talking about? Business planning? "Yes, I'm in total agreement with that last statement," I blurt out, swiftly shutting my half-dozen open browser tabs and taking a deep breath. Did everyone know I hadn't been listening? "Oh, wonderful, that's good news," someone says. "Thanks, Erica." Saved. Barely.

We all have weaknesses, and mine are browser tabs and mute buttons. As everyone knows, it's dangerously easy to multitask during a phone or conference call, though I'm guiltily aware it reduces my active listening. At least I'm not alone. In one study, roughly 65 percent of respondents admitted to doing other work or sending emails while participating in conference calls.[9]

That's why I've made it a point to ban the use of the mute tool on all my team calls. I also try to plan meetings that are to-the-point and engaging so that the participants are less tempted to let their minds wander.

I once presented a workshop before 30 people at a pharmaceutical company. Everyone in the audience seemed engaged, with the exception of one woman in the back row who couldn't tear her eyes away from her phone. Even when I was standing three feet away from her, she remained glued. It bugged me, and it even distracted the other people in the audience. And she was one of the most senior people in the room! We've all done this. Our goal, then, is to become aware of the perils of multitasking and realize how it affects our *own* attention.

• • • • • •

In the end, the goal of Value Visibly is very simple. It's all about making people feel appreciated in the workplace. Use the techniques in this chapter to ensure that you are consciously valuing your team online. Use the following assessment to analyze whether Value Visibly is present on your team. Check one box next to each statement. The more you "Strongly Agree," the higher the Value Visibly level is in your organization.

	Strongly Agree	Somewhat Agree	Somewhat Disagree	Strongly Disagree
Excellent work is acknowledged and rewarded in your organization.				
Your expertise and skills are valued and deployed.				
Your time is respected.				
You are not overworked or burnt out.				
Get the full assessment at ericadhawan.com/digitalbodylanguage				

●

CHAPTER 5

Communicate Carefully

Think Before You Type

We **Communicate Carefully** by sending messages that say what we mean and state what we need—from whom and when—thereby eliminating frustrating ambiguity across teams.

As the youngest child in an immigrant Indian family, I picked up basic English grammar fairly easily. But I still lacked many of the contextual cues that came naturally to my peers. I remember once inviting a middle-school friend to join my family for dinner at a local restaurant. At one point my friend whispered to me that the waiters considered our party "rude." It wasn't what anyone said, it was our tone and our cadence. You see, in Indian English, when people ask for something, they use an intonation with a falling contour so it comes off sounding like a statement, rather than a question. Most Americans are accustomed to requests that end in a rising contour. At that moment I knew exactly what my friend meant: without being aware of it, everyone in my family came off like hosts ordering around their staff!

When people communicate, without being aware of it, they make use of a broad range of "contextualization" cues that help others assess the meanings behind their words. For instance, the phrase "I love that film" accompanied by a head nod signals something entirely different from "I love that film" paired with an eye roll or a wink.

As I noted earlier, we're all "immigrants" in today's digital workplace, meaning that the subtle cues that help us understand what others may *really* be saying require time, patience, and even reflection.

Consider, for example, the story of Docstoc, an online document-sharing resource that launched in 2007. On the first day of its launch, Docstoc attracted 30,000 unique users.[1] CTO Alon Shwartz found this number to be cause for celebration. Thirty thousand unique users! But when he shared this number with the company CEO, Jason Nazar, his enthusiasm was quickly doused. Here is their conversation, in summary:

> Shwartz: We got 30,000. That's great!
> Nazar: We got 30,000? That's horrible!

Shwartz's success, it seemed, was Nazar's failure—and they were working on the same project! (Older readers will no doubt recall the scene in the 1977 film *Annie Hall* when a therapist asks the protagonist, Alvy Singer, how often he sleeps with his girlfriend. "Hardly ever," he replies. "Maybe three times a week." Asked the same question, Annie, his girlfriend, answers, "Constantly. I'd say three times a week.")

Ultimately, Shwartz and Nazar realized that they'd never taken the time to define what success looked like. Shwartz's takeaway? "If you don't define what success looks like, and you are not engaging in mutual validation, how do you know when you've succeeded?" He adds, "It's hard to reach a goal that is not clearly defined."[2]

Stories like these are pretty common. The complaints range from "Our departments have no common language" to "Nobody knows what our division is up to." In the end, it all comes down to a major stumbling block: no one is Communicating Carefully.

Up to 80 percent of all projects suffer from a lack of clarity and detail.[3] A recent survey shows that 56 percent of strategic projects fail as a result of poor communication.[4] In the United States alone, this adds up to a loss of $75 million for every $1 billion spent.[5]

When I met with my client Selena, she was nearly at the end of her rope. In a new role as a manager of a design team, she was busy building relationships with a team spread across the Eastern Seaboard. Problem was, she'd been clashing with one of her senior designers, who had sent her what she considered a sloppy, incomplete second-round draft. *This is ok*, Selena wrote back, enclosing a list of the things she thought worked, and that still needed fixing. *Ok. I'll get the changes to you shortly*, came the reply. But when the next draft arrived, none of the changes Selena had requested had been made, prompting her to call him and *command* he make the changes. The designer was incensed. *You told me that this was ok. Now you're lashing out at me?*

Lacking cues like eye contact, tone of voice, or body language to clarify Selena's intent, the designer interpreted her *This is ok* as more or less straightforward feedback. It wasn't—it was a subtle warning. The same went for the changes Selena proposed—they came across as *optional*. When the designer ignored them, was it any wonder Selena felt he wasn't listening or that he believed his boss was now changing her mind?

Gently, I explained to Selena that the problem was on *her* end. Her natural mode of engaging with colleagues was well intentioned but off the mark in a digital age. If she wanted to succeed in her new leadership role, she would need to be much clearer. It wasn't that she lacked people skills—Selena was doing everything possible to connect with her managers and staff—it was that she needed to adjust her digital body language.

The good news? Selena realized that clarity trumped politeness and would also help her colleagues thrive. She changed her feedback style to be more direct—even including a bullet-pointed list of requests. In no time at all, the designer delivered on what Selena wanted. It was all good.

Once, people could "clarify" what others were saying by picking up on their physical reactions—a quizzical expression, a stunned stare, the hint of a smile. This still works in real life, just not in the contemporary digital world. Even on video calls, there is a disconnect when you can't tell if people are looking at their cameras or not, or you can't quite gauge their expressions in the small participant boxes. Today, it's everyone's responsibility to consider the potential ways our communications can be interpreted (or misinterpreted) and adjust our writing style and tone accordingly.

Communicating Carefully means putting out clear signals that keep everyone fully informed and aligned. It doesn't mean that everyone has to agree—which almost never happens—but it does mean that goals are understood and shared. When team members are genuinely aligned on objectives and expectations, this higher level of mutual understanding frees everyone up to focus on being the best at what they do.

....

Communicating Carefully means getting to the point while considering context, medium, and audience.

....

The speed and pace of change in most businesses today make Communicating Carefully even harder to implement. Company leaders used to spend months crafting a robust vision of strategic alignment before carefully communicating that vision over a face-to-face campaign aimed at investors, business units, and customers. Today, leaders need to disseminate information *fast*.

This means that people are expected to present their ideas in bullet points, using headlines as proxies for the ideas that support them. The universal chaos of unread emails, IMs, texts, and calendar invites makes most of us crave a simpler, easier era of phone calls, office drop-ins, and uninterrupted client dinners. Back then, a day could go by before we finally responded to a voicemail (*Call Jack back*, we'd scrawl on a random piece of paper, which we'd then lose). Needless to say, Communicating Carefully in today's high-velocity, shorthand world demands a more concrete approach.

COMMUNICATE CAREFULLY: THE PRINCIPLES

Since many of us communicate most of the time with our thumbs, we need new *rules* of thumb to help us communicate clearly and persuasively.

Think Before You Type

It was 8:00 p.m. on a Sunday night, and I was frazzled and tired. Unfortunately, I couldn't start winding down before sending a few emails in anticipation of a busy Monday morning involving meetings and travel. Wearily, I drafted an email to my client Katie, incorporating specific guidance connected to her current team challenges, along with a draft of a presentation I would be sharing with them later that week. It looked good, I thought: clearly written, with bold headings, bullet points, and italics. I typed *Katie* in the To line, and when her email address popped up, I pressed *Send. Done.*

Two seconds later, my relief gave way to panic. I'd sent my email to *another* Katie at *another* company. This "other Katie" was a potential business prospect, a woman I was hoping to work with someday. I felt embarrassed—how stupid this must look to her. If I had only taken a few seconds to think clearly and be more careful about my communication, I could have avoided the whole thing.

It may seem obvious, but this kind of thing happens all of the time. And once we press *Send*, we cede control over where our words end up. A private email we send to an acquaintance might show up later in a post on his or her public Facebook page. Messages and posts can be copied, forwarded, altered, and updated in ways that distort their fundamental meaning, not to mention translated instantly (and not always correctly) into almost any language. An email can show up from a customer or client without us knowing that our boss's boss is included as a Bcc.

All this means one thing: we need to be *very careful.*

One former manufacturing executive I know sent his colleagues a twelve-paragraph email giving them a friendly heads-up about a possible future acquisition. Without his knowledge, two words from his long email were copied (out of context too) and widely forwarded across the organization: *Expect layoffs.*

A hospital administrator experienced the embarrassing and painful chaos that Reply All can create when she emailed the entire hospital staff with the latest draft of a controversial policy. Despite her best efforts to regain control, she spent the next week fielding Reply All feedback from 800 or so staff members.

Our fast-moving culture means that we don't always take the time we need to proofread or truly consider the words we've written before we press *Send*. But today, re-reading an email before we send it (and not ten times *after* we send it) is *mandatory*. How often have you heard the words "But I sent you an email" or "Didn't you get the email?" or "I am sure I covered that in the email" when you're staring directly at the email in question and the information just isn't there?

The first rule of Communicating Carefully? *Slow down.*

TRY THIS THINK-BEFORE-YOU-TYPE CHECKLIST

- Who needs to be included on this message?
- What do I want the receiver(s) to do after they read this message?
- What context or information do they need?
- What is the appropriate tone?
- When is the best time to send this message?
- What is the best channel to convey this message?
- How comfortable would I be if this message is screenshotted, forwarded, or otherwise shared? What can I do to change it? Or should I save this for a phone call or face-to-face meeting?

If you don't take a few moments to slow down, consider the following, very common scenario as something that may happen. Roz sends an email to her colleague Jon:

Hi, Jon, How was your ski trip this weekend? Hey, can you possibly send me the sales recap that went around to your group on Friday? I need to put a sales report together for the team. Also, did any of the accounts fail to report their numbers? Thanks so much for your help!—Roz

Seconds later, Jon responds via text:

Yep, we did $457K for Feb!

Always eager to please, Jon feels great about the exchange (fact is, he barely thought about it afterward). He's helped Roz out, and now he can move on to the next agenda item. A win-win. On her end, Roz is irritated for several reasons. First, she needed the *full* sales report; second, Jon didn't answer her question; third, Roz sent an email, but Jon answered her with a text (*why?*); and finally, Jon skipped over the pleasantries, not even acknowledging Roz's folksy greeting. (Ironically, Roz would have felt annoyed if Jon had taken *too much* time to respond.)

It's easy to assume that suspending cognition before typing is a "younger person" thing, common to so-called digital natives. But my research has shown that perpetrators of this particular crime of digital body language come in all ages and occupy all levels in organizations, including (and especially) executives, who are responsible for communicating so many messages that most settle for speed over clarity. The result is widespread confusion across teams.

One former client of mine, Joe, had a team with a low employee engagement score. As we dug in to figure out why, we saw that many of his team members reported feeling overwhelmed. They often worked on weekends and late at night. Joe realized two things: he hadn't created time boundaries around when his teams should communicate, and he was also guilty of sending emails at all hours. Was it any surprise his teams felt they needed to be in touch 24/7?

So he made some changes. Today his teams know that Joe writes and answers emails on weekend afternoons but doesn't expect a response until Monday morning. Joe even went the extra step of coining a new acronym, ROM, or "Respond on Monday." This way, Joe doesn't have to wait to send the email (and risk forgetting it altogether), but his team still gets their weekend.

Be Tone-*Deft*, not Tone-Deaf

Tone—the overall attitude, or character, of a piece of writing— is another key component of Communicating Carefully. Ask yourself: Who is the recipient? Who is the audience? What's the context here? Tailor your communication accordingly, or as I tell my clients, make sure you "read the room."

Naturally, this means anticipating how your words are likely to come across to others. When you write, text, or call your boss or colleagues, for instance, it's best to keep your tone neutral until you develop a rapport that would indicate differently. Focus on being informative, or persuasive. Edit yourself so that you stick to the essential facts.

Offline, a loud tone of voice can convey emphasis (*This matters!*), serve as a switching signal (actually, *this* is the thing that matters), or express extreme feelings (*I'm furious!*). A softer voice conveys "I don't know," signals calm, or indicates that maybe it's time for someone else to speak. The good news is that you can also modify the volume of your *digital* voice.

Consider the following email message: *THIS IS NOT GOOD, NEEDS A LOT OF WORK!!!!* It sounds like Zeus ordering a hit job on a lesser god—all caps, terse sentence structure, and a crazy picket fence of exclamation marks. If someone was trying to tear your head off, then mission accomplished. But if that same person was trying to convey respect, whoops!

Be aware, then, of the *visual* impact of your message.

Ethan, a manager I coach, once told me about an interaction with a senior leader that left him feeling unappreciated and belittled. As requested, he had sent this leader a detailed plan about increasing productivity. The plan set forth a different way of working that Ethan was certain could help teams avoid duplicating their efforts while creating new levels of transparency. Ethan was excited about the plan, and even included specific questions for the next team meeting. Expecting a positive response, maybe even a few follow-up questions, what he got back from the executive was this: *k.*

Sorry, *what*? K-what? K-pop? Ethan felt confused and insulted. His proposal was clear and comprehensive. Didn't it invite a matching response? Was the executive even *thinking* about Ethan's plan—or was she dismissing it outright? Did *k* mean she was giving him the green light to proceed, or was it a subtle command to put his dumb idea on a back burner? It was impossible to tell. Also, did the senior leader think so little of Ethan that she couldn't be bothered to write more than a single letter? Even something pedestrian, like *okay I'll get back to you*, would have conveyed more respect and attentiveness than that *k.*

HOW DO I WRITE MESSAGES PEOPLE WILL ACTUALLY UNDERSTAND?

- Assign email tasks using the "3 Ws." Every message should have a clear Who (the name of a specific person instead of a group); a clear What (an explicit description); and a clear When (the exact time and date, 4H = 4-hour deadline, 2D = 2-day deadline).

- Specify in the subject line or first sentence of a message if it's an FYI, a Decision Request, or a Request for Information.

- Create clear acronyms (NNTR = No Need To Respond; WINFY = What I Need From You).

- Write the perfect subject line. Summarize the body of your email, use prefix modifiers, and stick to the same subject line unless the subject changes (this isn't the place to throw in new or tangential questions). If the subject does change, create a new email thread with its own subject line.

- Break long messages into two parts. Label them "Quick Summary" and "Details." Know what you really want first, then get to that point at the top of the email.

- Make your messages scannable. Use bullet points, subheadings, white space, highlights, and bold text.

- Show instead of tell by attaching screenshots. They are valuable when you need to give someone instructions, or to highlight slides in a deck.

- Use if/then statements to increase accountability, create expectations, and provide clarity on next steps.

- Present options: ask, "Do you think we should do A, B, or C?" Try not to ask open-ended questions like "What do you think about this?" or "Thoughts?"

Tone deficiencies go well beyond team discussions. Some can come close to blowing up a brand.

The first week of April 2017 didn't start out well for United Airlines CEO Oscar Munoz, and in no time at all it turned into a disaster. A viral video emerged showing a United Airlines passenger being yanked from his seat and dragged down the aisle of an airplane by a team of security officers. A terrorist event averted? Uh, no. United had overbooked the flight, and when a randomly selected passenger refused to vacate his seat, the man, a physician, was forcibly removed.[6]

Obviously, this required a rapid response from the company and its CEO. United representatives, however, tweeted a tepid apology for having overbooked the flight without even mentioning the passenger, a tweet that was unanimously derided and ridiculed. Six hours later, Munoz tweeted his own public apology, one that was widely received as lame, tone-deaf, and the *opposite* of saying, "I'm sorry, we messed up."

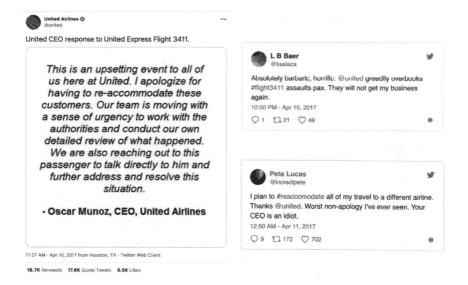

While Munoz and United eventually issued a proper apology, the damage was already done. Technically, of course, Munoz did the right thing by responding personally. Nonetheless, his tone was detached, and his language sounded hollow and inauthentic. The tweet also came too late. The world's angry, derisive response to how United mishandled a public relations crisis resulted in a temporary $1.4 billion drop in the airline's stock price.[7]

It didn't have to be that way. In 2018, Southwest Airlines came up against their own terrifying situation when an engine exploded midflight, obliterating an aircraft window and killing a passenger.[8] When the flight landed, Southwest responded quickly.

First, the airline released a short announcement on several social networks, reporting everything they knew. A follow-up announcement provided a link to an appropriately emotional in-depth statement issued by Southwest:

> We are deeply saddened to confirm that there is one fatality resulting from this accident. The entire Southwest Airlines family is devastated and extends its deepest, heartfelt sympathy to the customers, employees, family members and loved ones affected by this tragic event.[9]

As the airline amassed additional facts, they passed them to the world in real time, along with a video from company CEO, Gary Kelly.[10] Southwest's website header and Twitter profile image changed overnight from a default red, yellow, and blue heart symbol to a gray shattered heart. All other marketing messages were pulled. It was obvious that the Southwest PR team had considered the broader context, put themselves in the public's shoes—and struck the right tone.

KNOW WHEN TO CHANGE THE CHANNEL

Several years ago, following a collaboration workshop I led at a large retailer, I was asked to work with two employees whom

I'll call Samantha and Tony, who had agreed to act as "collaboration champions" despite living in different US states and communicating primarily via email. Their job, we agreed, was to continue to model and reinforce various behaviors we'd discussed during the workshop.

Well, if Samantha and Tony were collaboration champions, I would hate to see who came in last. The complaints started almost immediately. According to Samantha, Tony was a smug, annoying smartass. I tried to stay above the fray as I gave Samantha tips on how best to deal with Tony's particular brand of sarcastic humor. At the same time, I was also counseling Tony about putting aside his sarcasm in favor of optimal clarity.

Still, nothing I did seemed to work, and the tension between the two only worsened. A month later, Samantha had had enough. She wrote a forceful, aggressive email to Tony detailing how she hated his tone, how his sarcasm was driving her nuts, and why, in her opinion, he wasn't taking the project seriously. The worst part? She cc'd me and affixed this note to the end of her email:

Erica, I believe you agree with me.

That wasn't a good day. I found myself caught in a back-and-forth email chain, all the while feeling like a put-upon preschool teacher on the verge of screaming, "Time out!" In an effort to hash things out between them, I set up a video conference call.

Ultimately, Samantha and Tony were able to tolerate each other—just barely, but enough to complete their tenure as—what was the hopeful expression?—"collaboration champions." It wasn't that they couldn't work together. Instead, they needed to change the channel they were using from email to regular video calls. Video calls eliminated the miscommunication that resulted from their wildly different email styles while adding visual cues that conveyed their emotions and (it's true) mutual good intentions.

Then there are some people who get it right from the start.

A web applications company called Zapier has seemingly cracked the code for using the right collaboration channels, even coming out with a guide to help other companies do the same thing.[11] Zapier's workforce is entirely remote, with almost everything shared via written communication. Every one of the company's collaboration tools serves a clear-cut purpose while mimicking a parallel real-world office environment.

For example, the instant messaging platform Slack serves as Zapier's virtual office, i.e., "if you're in Slack, then you're at work."[12] Within Slack, company employees have created both work- and non-work-related channels (messaging chains), ranging from "marketing" and "hacking" to "water cooler." These discrete "rooms" ensure that employees' messages are seen by the appropriate audience, especially as teams grow beyond a dozen or so people, decreasing the likelihood of duplication across channels.[13] Every "room" has its own working policies that are revised and upgraded over time as well as its own moderators. Zapier also makes use of Slack integration tools like Trello, GitHub, and Google Docs, with each one, again, serving a clear purpose. Finally, the company keeps everything organized by adhering to detailed norms that specify when and how to use each tool.

One of the most powerful things we can do is take pains to specify appropriate channels and formats across the culture. You don't have to always stick to the same communication medium, but make sure you choose the *right* one for the message. Using the wrong channel at the wrong time can even have professional consequences, as it could damage trust in you or even brand you as unsophisticated or unsympathetic.

I teach clients about the implications of choosing the right channel by focusing on three factors: **Length, Complexity, and Familiarity**.

LENGTH, COMPLEXITY, AND FAMILIARITY

Length

Of these three factors, *length* is the easiest one to manage. Most of us have family members or colleagues who send multiple, consecutive, lengthy texts, as they're seemingly unable to squeeze their thoughts or ideas into one or two. If you want to provide someone with a lengthy (e.g. more than a short paragraph) update, use email, and for heaven's sake, refrain from using channels like instant messaging. Use bold, underlined headings, include attachments if applicable, and make sure you briefly share specific context up front so readers understand why it's important. Bottom line: if your message doesn't match the medium, find a more appropriate channel.

Complexity

Complexity is a lot harder to figure out. The general rule of thumb is that bigger, broader ideas require more reflection and/or nuanced thinking. If you're gearing up to make a complex argument, it's best to choose a medium (including a deck or a blog) that permits a greater level of detail and also supports add-on elements like photos, videos, or space for feedback or comments.

Again, always be aware of the *visual* impact of your message. If it goes on too long, you risk overwhelming people. Also, too many boldfaced or underlined words can come across as chaotic. Images (appropriate ones) can both clarify and enhance trust, but superfluous ones are too often distracting.

Be conscious as well of the *timing* of complex messages. Don't expect your team to absorb a novella-length email that you decide to send out at 5:00 p.m. on a Friday, and respond to it thoughtfully an hour later. Nor should you wait until the last minute to send complex messages that may invite some back-and-forth. In both cases, it's best to end your message with an invitation for a phone call or an in-person meeting to go over the smaller details.

Again, always use the channel that suits the tone and message you want to convey, bearing in mind that just because we live in a digital world doesn't mean digital is the only way to go.

Complexity of Argument	Best Channels to Use	Why?
High	Articles, blogs, visual presentations, video calls	Allows for greater trust-building; can include supporting elements like photos, videos, and a feedback or commenting option
Medium	Email, phone, and group conference calls	Allows for context and back-and-forth discussion
Low	Text, IM, and group chats	Allows for quick responses and requires less context

Familiarity

Familiarity refers not only to our relationship with the recipient(s) of what we write but also to the *content* of what we're saying. Who is your audience? If you have a close relationship with someone, sending that person a text may be a welcome,

COMMUNICATE CAREFULLY O 121

neutral disruption. But in a business relationship, most people prefer communicating via email, which allows them to scan the subject line and decide when (or even if) to open and read the message.

Consider your content. Is it personal, confidential? If it is, make sure you build trust by sending a direct private message instead of a public group IM.

LENGTH, COMPLEXITY, AND FAMILIARITY: PUTTING IT ALL TOGETHER

We've now gone over the three factors involved in choosing the channel that best shows your respect for other people's limited time and attention. With each one, of course, there will be variations or special circumstances. For instance, I often hear that video meetings are a preferred option when face-to-face meetings aren't possible. But as we all know by now, video meetings are far from perfect, with many feeling decidedly *un*collaborative and more like siloed lectures where one or two people at a time bang on while everyone else is forced to listen. But there are ways around this. Zoom, for example, offers "Breakout Rooms" that sequester meeting participants in standalone rooms of any size. Zoom also lets participants use a virtual whiteboard during sessions, so that everyone present can write and collaborate simultaneously, thereby building shared context.

Recently, an organization brought me in to assess one of its team's digital communications, which made up most of that team's "update culture." The division leader wanted to know why there was so much daily dysfunction: missed deadlines, ignored emails, uncomfortable chat room conversations, and pervasive peer-based passive-aggressiveness.

It didn't take me long to discover that the team in question was using its collaboration tools in every which way but the right one! In the team's hands, Microsoft Teams chat had become a way for members to avoid face-to-face collaboration. Members were also sharing messages and documents using multiple collaboration tools at random, making it hard for anybody to know where to go for what. Finally, some members were commenting on tasks using ten-word IM messages without explaining if their response was an opinion or a request for action.

Eventually, the team and I created norms around the best, most proper use of every communication channel. Here's what we built:

Tool	When to Use	Response Time	Norms (how to use and not use)
Skype Messenger (In-House IM)	Time-sensitive, urgent messaging Short and simple conversations	ASAP	Use with less than 6 people (otherwise call) Always set your own availability on Skype Avoid complicated questions or conversations that require visuals
Email	Provide directional, important, and timely information Ensure there's a record of your communication Direct the receiver to an online source for more information	<24 hrs; priority dependent	Use identifiers in subject line for urgency & response expectation Use to share attachments Avoid when immediate response is required Not for random chit-chat
Video Call	Use for meetings, including external ones that could	Schedule in advance; priority dependent	Ensure appropriate usage of camera & mic

	benefit from visual interaction (e.g., project check-ins, introductions, deck sharing)		Use "mute" when needed Ensure meeting host clarifies if video functionality is required for participation Record calls for those who miss them
Texting (Cell Phone)	Time sensitive/ urgent communications Only use if you were unable to reach via other channels	Within 30 minutes if between 7am and 7pm; priority dependent	Use can be adjusted if it is the preferred communication for your leader Avoid texting during meetings/working sessions

Setting collaboration channel norms wasn't hard. No, the hard part was making sure that teams *followed through* with these new behaviors and didn't gradually slide back to their old ways. Mindful of this, we identified two or three channel advocates whose role was to encourage best practices within each channel and give shout-outs to those who were modeling the right behaviors. Finally, we developed a practice designed to eliminate situations where individuals duplicated content unnecessarily across multiple channels by rolling out the hashtag #killduplication.

The #killduplication phrase is now a staple in the team culture, helping to eliminate wasted time and ensure colleagues optimize the use of each digital medium.

THE COMMITMENTS ARE IN THE DETAILS

A long-standing client of mine, an accounting firm, once called me in the wake of a failed product launch targeted at a new customer segment.

The project had been launched only a few months earlier. Here's how it was done. First, the marketing department invited 660 attendees (50 percent of all employees) to a town hall meeting hosted by the CEO, who presented a formal PowerPoint presentation followed by a half-hour Q&A. Other members of the marketing team hosted a follow-up chat on Yammer (a closed-network social platform) and a special lunch for second-level managers. Everyone went around using language like "This is a *big* opportunity" and "Let's spark a change" to make sure employees understood the extraordinary nature of this new project. They were perplexed when, a month later, nothing happened. All of the ideas had become stuck in the planning stage, and many were set aside completely for "more pressing" work.

Well, here's *why* nothing happened. At no time did the leadership team ask managers to write down clear, measurable commitments around what they and their teams planned on doing to engage customers, work alongside the marketing department, or prioritize benchmarks.

To get everyone back on track, the executive team and I asked each team member to communicate the details of the new offering to existing and potential clients by the end of the week. Next, we used Microsoft Teams to track their progress, setting mechanisms in place to provide scripts, share client FAQs, and enlist executives in high-priority discussions. This simple practice allowed the teams to make fewer, better commitments and to stick to them.

The hardest part of any promise? *Staying on track.* The more networked an organization is, the more on track a team has to be to deal with touchpoints and requests. To reduce confusion, I always advise writing down and tracking individual and team commitments. At every stage of a project, list your key ongoing commitments. For example, "I will provide my team with the resources they need in these areas: website updates, product rollouts, and communications." Under those, list your specific promises, such as "I will hire two new engineers by the end of September."

HOW CAN I MAKE SURE MY TEAM FOLLOWS THROUGH WITH COMMITMENTS?

- Is the commitment observable and measurable?
- If the commitment is carried out well, will it support or leverage change that works toward the team's stated goal or benchmark?
- Do individuals or teams have all of the resources they need to fulfill this commitment?
- Does this commitment risk overextending the team? Will the team require additional support?

Weak Commitments	Powerful Commitments
At the next team meeting, I will share what we talked about at the offsite.	I will meet with my reports next week to review our revenue projection and complete a validation analysis. This information will inform the next steps we'll take to meet our targets.
I will talk with the director about the need to improve outcomes.	I will collect information and data to demonstrate what is and isn't working as we try to change culture. I will link to best practices from other companies. By the end of the month, I will make a culture implementation recommendation to the director.
I will design a recruiting event.	By next month, two Midwest recruiters and I will attend two state job fair conferences to gather best practices on designing our events. We will present our design recommendations at the next quarterly meeting of the talent group.

I will use performance feedback in my work.	I will review my feedback and reflect on the conflict that arose between my director and myself. I will implement a strategy to address that conflict by next week, with a plan on how to address each of his possible reactions.

Communicating Carefully establishes the base from which a team can successfully execute its goals. Ask yourself, "What do I want the person who reads this *to do* after I communicate this message?" By considering the Who, What, When, Where, and How of all your communications and including the context the recipients of these messages might need to understand them more fully, your team's performance is bound to improve.

Also: make sure you measure success with *details*. Confirm that everyone understands what he or she has agreed to, including the owners, actions, and deadlines you expect. Lastly, set up a process to regularly review these measures of success to track progress and make ongoing adjustments whenever necessary.

Put simply, Communicating Carefully is about ensuring that all the relevant players are on the same page. It's harder to accomplish this when you're working with digital teams, but not impossible.

Remember: *think* before you type, choose the *right* channel, and focus on the *detail* of your communication.

Use the following assessment to analyze whether Communicate Carefully is present on your team. Simply check the box beside each statement that most closely resembles your level of agreement with the statement. The more "Strongly Agree" boxes are checked, the higher the preexisting level of Communicating Carefully is in your organization.

	Strongly Agree	Somewhat Agree	Somewhat Disagree	Strongly Disagree
You understand the specific goals and objectives of each team project.				
After each meeting, you are clear on next steps and have minutes to look back on just in case.				
Your team has a clear set of norms regarding channel selection and response time.				
You understand what is being asked of you in the messages you receive.				
Get the full assessment at ericadhawan.com/digitalbodylanguage				

•

CHAPTER 6

Collaborate Confidently

Teamwork in a Digital Age

My Indian family is full of loud talkers. It's just how we were raised, how we learned dialect in our culture. In my family, either you talk loudly or you don't get heard. As a public speaker, I don't even need a microphone sometimes, even when I'm talking to 100 people or more.

Still, it wasn't until I was sharing an office with a too-loud colleague that I realized the downside of my own volume. His glass-enclosed office space was directly across the hallway from mine, yet it always seemed like he was screaming directly into my ear. I couldn't hear myself *think* around this person, but I was afraid to say anything. Was there a polite way to tell him he should consider lowering his voice during meetings and phone calls? If I shushed him, would it offend him and affect our ability to work together?

I rehearsed a few different ways of saying something. *Should I tell him over lunch, casually at the watercooler, or as I'm walking past his door? Wait! Should I get someone who knows him better to do it? No, wait again! I'll write him an email!* I drafted one, slept on it, and deleted it the next morning.

Later that same day, I said to him finally, "Could you please speak a bit more quietly or maybe close your door when you're on calls?"

"Sure, I'm so sorry," he said.

That was it. It wasn't hard at all. Imagine all the time and energy I could have saved if I'd been confident that my request would be taken as helpful and well intentioned (as it was).

Confident collaboration requires us to put aside what-ever fears and anxieties we might have about what others might think or say and speak up! I'm sure we've all had a colleague who needs everything *right now*. All his emails have *URGENT!* in their subject line, and if that weren't emphatic enough, he texts you a few seconds after emailing, and then, if you haven't responded in a few hours, the phone rings, and guess who it is? You might have even dropped a deadline or two of your own to help him finish something—he's on an important deadline!—only to discover that his deadline was in his own head.

Or maybe you know his counterpart—she has the opposite problem. She's habitually *un*-urgent. She agrees to complete a large piece of work, but when you send her multiple follow-up emails, she doesn't answer, even when the deadline has come and gone. Is she stressed out, or overburdened? You have no idea, but as you await a response that never comes, you end up missing your *own* deadlines.

People like these make collaboration difficult, especially in a digital workplace. Often, negative workplace behaviors like these are based on fear and anxiety that mutate into chronic delays, passive-aggressiveness, and the erosion of trust.

Based on responses from a CEB workforce survey of over 23,000 employees, a recent *Fortune* study reported that 60 percent of all employees have to consult with at least *10* colleagues daily just to get their jobs done.[1] Half of that same 60 percent needs to engage with more than *20* colleagues to do their work.[2] Over the past five years, the time required for one company to sell something to another has risen 22 percent.[3] In a world entrenched in teamwork, we need to focus on getting beyond negative, fear-based workplace behaviors to uncover the best ways to collaborate.

We Collaborate Confidently when we state our needs clearly, including *when* and *why* we need something, leaving no room for misinterpretation (or fear, or anxiety).

Why is it so hard to Collaborate Confidently in the digital workplace?

In traditional office settings, it was easy to pop by a colleague's desk for short conversations—"Do you have a minute?"—or share a meaningful look across the room. These were once ingredients of a larger culture in which our social relationships at work served to strengthen trust and mutual understanding. This spontaneous, drop-in quality is largely missing in today's work environment. In some cases, we never get to meet our colleagues face-to-face. We may even have dozens of them, spread across different departments and time zones.

Team members are more likely to say something like, "Sorry, saw you called, didn't listen to the voicemail, could you send a video meeting invite instead?" or they may flat-out tell you they're just too busy to schedule a meeting. Thoughtfulness, we hardly knew ye. Everything has to be done *this second*, and the fact is, most people's brains—and schedules—just don't work that way.

COLLABORATE CONFIDENTLY: THE PRINCIPLES

Stay in the Loop

Collaborating Confidently is about keeping all relevant parties informed and up-to-date while checking in constantly to ensure ongoing clarity in *all* components.

For example, one of my clients, Kerry, the COO of a division in a technology company, recalls an example of misalignment among project components: "I had to update a senior executive with a project plan that involved three teams in my division. All three teams knew we had to send him one plan. All the different teams sent me different timelines for the final deck. They didn't even talk to each other about aligning schedules. I'm inundated with insufficient information, it's 6:00 p.m.— and the senior executive expects it in his inbox by midnight."

Or here's another. An American company in the consumer market was planning a rollout across Europe of a new personal hygiene product. Françoise Henderson, a translator on the team, was working on various marketing materials. At one point she noticed that the product's ingredients list included in advertisements was different from the list appearing on the bottles themselves (some ingredients, it seemed, were banned in Europe).

"No one told the marketing department, though," Henderson said.[4] Five separate company units—marketing, communications, technical, legal, and packaging—were supposed to be apprised of updates and changes throughout the rollout. In this case, no one was on the same page.

> Collaborating Confidently begins by understanding what other departments do— and establishing clear norms on how they interact with one another.

If anyone understands the problem of confidence gaps in collaboration, it's Lisa Shalett, former partner and head of Brand Marketing and Digital Strategy at Goldman Sachs, who now serves on corporate boards. She created a task force at Goldman Sachs comprised of employees and departments that were equipped to address a broad range of topics, including "legal, compliance, employment law, employee relations, technology, information security, [and] operational risk."[5] Why were there so many experts in the mix? "So we can get to yes faster," Shalett says. "Or, if we have to get to no, at least we all are comfortable with why we ended up at no."[6] At the beginning of any project, Shalett advises asking: "Who's going to need to know all the things we're thinking about doing? Where are the risks? Where are people going to really need to understand the processes, the requirements, the regulations?"[7] Most critically, she advocates identifying and prioritizing the right people who can make a project better, as well as those who can best anticipate bottlenecks, or call out the BS.

Posing the simple questions "Who could derail this?" and "Who will need to approve this?" can avoid leaving out important individuals who could later slow down your efforts, from a frontline cashier to a new customer representative to a risk manager. Shalett considers *all* stakeholders, beyond her immediate team, including people who may not be making the decision but who will be implementing it. Additionally, she asks all departments involved to break down proposals and issues into layman's terms. For example, engineering, management, and legal all speak different languages. To ensure these disparate voices understand one another, each is expected to express their ideas using clear, jargon-free language.

Caroline, a team leader at a pharmaceutical company, designates "project team members" and "project advisors." Team members are involved in decision-making and maintaining day-to-day activity, while project advisors provide expertise on specific subject matter and are only included on meeting summaries (keeping them in the loop) or in one-on-one conversations. Project team members who *can't* attend a meeting are responsible for appointing a proxy to make decisions for them.

Assigning these roles has reduced Caroline's 30-person brainstorming meetings into 6-person discussions. Things now get done much more quickly and efficiently.

HOW TO BUILD TEAM CONFIDENCE

- **Measure success in results, not hours.** Avoid hounding people over not working enough or working too many hours. If you tell them they need to work eight-hour days, but they know they can get the work done in five, they will spend more time getting sidetracked instead of getting the job done.

- **Set clear roles and expectations.** Tasks should be framed as steps to accomplish a common goal. Be clear on the direct line between the two. Typically I end all phone calls with a question: Who is doing *what*—and by *when*? That way the entire team is clear, and individuals feel a sense of accountability toward their peers as well as the team leader. As tasks are getting done, I like to use a project-tracking software like Trello to increase accountability.

- **Agree on what success looks like.** At the start of a project, ask three questions: What does *great* look like? What does *done* look like? What is out of scope? From there, work backward as a group to decide on a realistic deadline.

> • **Be available.** If team members can't reach you with a question about a task, they are more likely to lose steam. Think of all the hours that could potentially be wasted if you don't make yourself available to answer a five-minute question!

WIN WITH CONSISTENCY

· ·

To battle today's erosion of confidence at work often caused by changing priorities, it is critical to be consistent in our messages.

· ·

How can we best build and earn confidence through consistency?

The answer: **fight thoughtless deadlines; eliminate chronic cancellations;** and **practice patient responses.**

Fight Thoughtless Deadlines

The word "deadline" can be traced back to the American Civil War—who knew? Back then, prisoner-of-war camps had boundaries known as "dead-lines"—prisoners who crossed them would be killed.[8] In short, deadlines were *serious*.

In some places they still are. In a manufacturing plant, for example, a missed deadline can cause chaos for countless stakeholders along the supply chain. But in other settings, there is less at stake in missing a precise deadline, and deadlines are understood to be rough—calibrated for "noonish," "ASAP," or "first thing in the morning." These requests may or may not feel urgent. Particularly in creative industries that involve iteration and innovation, it's understood that deadlines may not be met if the idea just hasn't reached completion. The problem arises

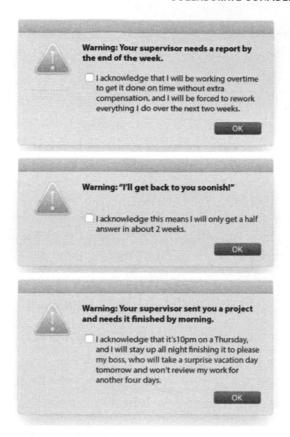

when one team or one person's missed deadline causes a delay down the chain for someone else.

Most organizations deal with all of these scenarios at one time or another. When people are collaborating from different places and time zones, observing disparate working hours, overcoming language barriers, and more, meeting deadlines becomes much more difficult for all. And so it's important for managers to have a system in place that creates realistic deadlines, clarifies the consequences of missing them, and considers contingencies for when (inevitably) something goes wrong.

Be thoughtful about deadlines. When setting major deadlines with her management team, Mary, an SVP at a hospitality company division, always prefaced her scheduling meetings with a caveat: "I'm thinking out loud and believe the deadline should be December 1. I haven't finalized the deadline. I want everyone to give me their opinion."

Why did Mary do this? To allow her team to speak up about any potential issues instead of remaining silent out of the mistaken belief that she'd already made up her mind. In addition, Mary *welcomed* email from employees who didn't speak up in meetings. Email gave those employees the psychological safety they needed to contribute to the conversation, helping them accept her final deadlines even when their opinions didn't change anything.

Eliminate Chronic Cancellations

Meeting Canceled Everyone- forgot I'll be on vacation so I'm canceling the meeting. See you in 2 weeks!		Close
Re: Re: Meeting Re-Resched... John, I can't make it at that time. I'll still be in the operations half-day meeting!		Close
Re: Meeting Re-Rescheduled JK, We'll be in conference room A instead. B will be occupied during that slot.		Close
Meeting Re-Rescheduled Sorry everyone. The Tueday meeting is now Thursday at 11am in conference room B.		Close

Canceling meetings is a real problem in the workplace. It's getting worse too, since we're all so overscheduled and overworked (at least we think we are). It's easy to book time on someone else's Outlook calendar—*why don't we just do it now, and bow out later if we have to?*

But chronically canceling meetings can have company-wide repercussions, including lowered morale, lost team brainstorming time, and a general loss of confidence in leadership.

Here's an example: Nadia runs an internal marketing team at a large insurance services firm. Her team worked three weeks on a one-year strategy plan for the chief innovation officer, only to have him cancel the meeting just hours before its start. Sure, the meeting was eventually rescheduled, but the cancellation (*all that work*) made the team feel devalued and unseen. It also signaled ambivalence—if the meeting were actually important, it would have happened.

Almost worse than the meeting getting canceled was the curt email announcing the news with no explanation. While you will inevitably have to cancel some meetings, when you do, there is a right way to do it. He should have sent a direct note saying *I'm sorry* and explaining why the meeting was canceled. He also should have been respectful in how he communicated his reason, such as *I understand how important this is . . .* or *Let's reschedule this as soon as possible . . .*

Practice Patient Responses

unsubscribe—a message sent to the massive Reply All chain that went too far

oops I really shouldn't email before having my coffee.—a self-correction to a hastily written, hastily sent email

Not trying to be rude, but . . .—the beginning of a passive-aggressive email sent a few beats too quickly

As I've said more than once in this book, today's channels are asynchronous, meaning that multiple messages can show up at the same time, throwing a wrench into the very concept of "sequence." We forget the truism that in the long run, *less* haste ultimately equals *more* speed. Missed or crisscrossing messages can undo collaboration confidence by creating misunderstanding, which leads to broken commitments or canceled meetings. They can also lead to widespread inaction or, worse, chaos.

The gaps we all experience in response times bring with them another problem, namely that circumstances can change dramatically before we've gotten an answer to the first email we sent. In addition, our need for a response increases with every second we sit there, making us feel impatient, resentful, and stressed out. This issue is especially germane for global leaders with teams in multiple time zones.

"I'll wake up in New York to a flurry of 50 email exchanges about an issue in our Shanghai office," says Sam, a shared services leader at Walt Disney Parks and Resorts. "People freak out because they haven't heard back from me yet, don't read the latest reply with an update before responding, and keep sending emails instead of giving me a call." As a manager of teams across his company's Orlando, Shanghai, and Paris theme parks, Sam goes bananas when his team sends volumes of emails without checking first that they're adding genuine value to the infinitely long Reply All chain.

If you can withstand the blind impulse to respond to an email immediately, well, there's a lot of power and control to be found in the ensuing silence. Unless the message is urgent or time-sensitive, don't drop everything to respond at once. The collective interest (your own included) is far better served by a measured, strategic reply. Silence allows us to gain perspective, to consider every angle, to review what's happened and anticipate what might come next.

Digital hastiness can also foster groupthink and undermine team creativity. Six *yes* emails in a chain make it harder for the seventh person to say *no*. A rushed "Does everyone agree?" at the end of a video meeting doesn't feel like a true invitation for discord.

Take a few additional moments of pause to re-read what you've just written. *Are you saying what you* think *you're saying?* For all its drawbacks, asynchronous communication gives us time to process our words instead of just blurting them out. Needless to say, this is a very real advantage. Don't automatically choose immediacy over a thoughtful response that can be all the more valuable.

HOW TO AVOID DIGITAL GROUPTHINK

- Be careful about saying yes to a task without first understanding all of the details, including what the deadlines are, who will be working alongside you, and what resources you'll have at your disposal. When these aren't clear at the start, communicate proactively to find out before planning a roadmap for the task.
- Don't expect instantaneous responses to the messages you send. Realize that sometimes people are working on other things. Set a reminder for yourself to follow up in two days—unless it's urgent, in which case you may try reaching out on a second channel once an unusual amount of time has passed.
- Re-read your messages at least once before sending them. Hastiness does not generate the best solutions.
- Check for both grammatical errors and clarity. Are you saying what you mean, providing necessary detail, and being clear in your ask? Do you describe what the deliverable should look like, the deadlines, and the check-in process?

FOLLOW UP STRATEGICALLY

In college, I had a roommate who, after every single face-to-face conversation, always liked to leave me a follow-up voicemail: "Just checking in if you were able to call the cleaning lady. The kitchen is really dirty. Also, wanted to check if you bought more paper towels. I think you said you were going to do that. Let me know." Needless to say, when I moved out, I felt like taking the paper towels with me.

Do you know anyone who actually *likes* getting an email whose first line is *Just checking in?* I don't! Nevertheless, the proper etiquette for following up is a critical component of collaboration. Are you supposed to follow up on a task using email, text, or phone? Is it okay to ask someone if they received the message you sent earlier that day, or does that come across as pestering and distrusting?

Collaborating Confidently means being able to follow up strategically—knowing when and how without qualms about it.

I NEED A RESPONSE. HOW DO I FOLLOW UP WITHOUT BEING A NAG?

- Amend the subject line to clarify that the email is a follow-up request and not a new task.
- Don't cc new people (unless you absolutely have to).
- Suggest another way to communicate (e.g., *Can we schedule a phone call?*).

EXECUTIVE PRESENCE IN A DIGITAL WORLD

The houses of confident leaders are always in order. Confident leaders never send emails scattered with typos. They don't forget to include relevant team members in group messages. They establish norms in their digital communications that create guidelines for their teams on how and when to communicate, what is appropriate behavior on each social channel, and more. Finally, they lead by example and follow these rules themselves.

....

Leaders with a strong executive presence are present, calculated, and careful. Online, this means double-checking all your written digital communications and treating virtual meetings as if you were there in person.

....

In digital meetings, executive presence makes itself known in obvious ways—the skill a leader shows by facilitating constructive discussions, for example, or the way she sidesteps the common pitfalls of the medium (people talking over or lecturing one another, or falling prey to offline distractions). Below you'll find a few more ways that leaders can show strong digital leadership.

First, understand that digital meetings demand more preparation than in-person meetings. Send out the brainstorm topic before the meeting time so that attendees can begin readying their contributions. By asking team members to bring their top three ideas to the meeting, you'll avoid creating incomplete solutions and running overtime. Consider asking your team to split into subgroups with their in-person colleagues to discuss ideas beforehand. That way, you can use the call time for ideas that have already been effectively pre-screened and pre-validated.

WHAT DOES EXECUTIVE PRESENCE LOOK LIKE ONLINE?

- You set deadlines in collaboration with your team.
- You send clear messages with a clear ask, not confusing brevity with clarity.
- You have a background on video calls that is not distracting to the group conversation.

- You acknowledge individual differences among team members and account for those needs.
- You set and enforce norms for team communications in a collaborative way.
- You serve as a facilitator, not a monopolizer, of team discussions.
- You are consistent in your words and your actions in a way that is authentic to you.

Even in a digital landscape, Collaborating Confidently involves creating durable commitments among your team. The key is stepping back and asking yourself what seemingly small things will lead to better collaboration. *Is it setting norms on the use of digital mediums? Is it responding to messages in full, and respecting others' time? Is it ensuring that your team's understanding of "success" isn't lost in an email chain?* Is it, quite possibly, *all* of these things?

Use the following assessment to analyze whether Collaborate Confidently is present in your workplace. Simply check the box beside each statement that most closely resembles your level of agreement with the statement. The more "Strongly Agree" boxes are checked, the higher the level of Collaborating Confidently is in your organization.

	Strongly Agree	Somewhat Agree	Somewhat Disagree	Strongly Disagree
Teams update one another regularly and follow up appropriately.				
Your manager or team leader is readily available for questions and support.				

	Strongly Agree	Somewhat Agree	Somewhat Disagree	Strongly Disagree
Deadlines are honored and valued.				
Members of your team always feel welcome to speak up if they disagree with a majority opinion.				
Get the full assessment at ericadhawan.com/digitalbodylanguage				

•

CHAPTER 7

Trust Totally

Innovate Faster and Further Together

You can't trust anyone! This perennial refrain echoed through-out my childhood. My parents, who had immigrated to Amer-ica so we could live a better life, also wanted to have a *safer* life. And one of the ways they taught their kids how to navigate the world was by instructing us to steer clear of dangerous things and people.

Don't answer the front door if it's someone you don't know. Stay off the five-rung ladder in the playground. Stop touching the dirt. And the big one, emphasized repeatedly: *Never, ever, ever get in a car with a stranger.* If anyone ever tried to force me to get into their car, my mother said, I should yell loudly and run away fast.

One day, when I was around 11 or 12, I had to get my teeth cleaned after school. The dentist's office was only a few blocks away, so once school ended, I grabbed my backpack and started walking there. A few minutes later, an older man driving past slowed his car, rolled down his window, and asked me if I needed a ride.

My mother's words flashed through my mind in red neon. I started running and didn't stop until I got to the dentist's office (never has anyone wanted to get to the dentist's as badly as I did that day).

I completely forgot about the incident until a few years ago. I was living and working in Boston, and since I didn't own a car at the time, I took the subway everywhere. One day, a friend told me how much time he was saving using this new thing called *Uber*. My mother's words floated through my brain. My friend was expecting me—*me*—to trust a stranger behind the wheel of an equally strange car? Then I found out that Uber's business model allowed passengers to access the driver's image and name, and vice versa. Uber would tell me how other passengers had rated the driver, and also how many minutes it took for them to show up on my street. I could even track our route.

So I scheduled my first Uber ride. It was exactly as advertised—safe, convenient, efficient. My trust increased over time, and the act of getting into a stranger's car slowly turned from an unimaginable idea into a twice-daily event.

I'm not saying that reaching a place where a team can Trust Totally is as easy or as snappy as hailing an Uber, but arguably the same elements—creating an environment of psychological safety where we're comfortable enough to speak up or take risks—apply. Creating a Trust Totally culture requires leaders who are willing to put in the work to foster psychological safety just as Uber did with its rating system, route-tracking functionality, and driver background checks. Ideally, Trusting Totally is what happens when we combine Valuing Visibly, Communicating Carefully, and Collaborating Confidently in practice.

Having said this, no one pillar can stand alone.

To wit: when we Value others Visibly on the path to Trusting Totally, we are showing them the same level of respect and acknowledgment we expect for ourselves. When Sarah stays up late to get her boss, Karen, a deck with a list of questions, Karen knows to send a quick email response: *Got it, thx and will review on Tuesday when back*. Karen knows that delaying her response until Tuesday would be disrespectful to Sarah, who's liable to assume something is wrong (when the fact of the matter is that Karen simply hasn't had a chance yet to review her work). By taking the time to send the message, Karen is ensuring that Sarah feels valued.

As you now know, Communicating Carefully is another building block that helps achieve a Trust Totally culture. Imagine the angst Sarah would feel if, after spending weeks working on her deck, Karen told her that the marketing department had come up with a different strategy, nullifying all of Sarah's deck recommendations. If Karen had only taken the time to align early on with the marketing team, she could have saved Sarah a lot of time and work.

Not least, a Trust Totally culture requires that, before anything, we Collaborate Confidently, eliminating the possibility of plans changing, or promises breaking, at the last second. You can Value Visibly and Communicate Carefully using an agreed-upon strategy, but everything will break down if Sarah's boss, Karen, is a mercurial leader slow to take action or make hard decisions. The same goes for teams without accountability measures, who miss deadlines or complete mediocre work without ever receiving constructive feedback. They need to know they can count on their leaders *and* one another.

WHY IT'S SO HARD TO TRUST TOTALLY

Some leaders say all the right things about "speaking up" and "taking risks," but many simply don't walk the walk, or support and empower team members who do.

"You could hear her typing and you knew something bad was going to happen," said one employee of her boss, the CEO of the luggage company Away. On the surface, Steph Korey seemed to have done everything right. By her late twenties, she'd gotten her MBA from Columbia and found a place at one of today's hottest startups, Warby Parker. As she and her friend (and former Warby Parker colleague) Jen Rubio brainstormed ideas one day, Rubio came up with what she thought was a winning concept: Luggage! Great luggage! Sexy luggage! Together Korey and Rubio conducted surveys to test the idea, quickly raised $150,000, and soon found themselves in China overseeing the first prototypes of Away luggage. The company's first suitcases came to market in early 2016, and two years later Korey and Rubio made both *Forbes*'s Next Billion-Dollar Startups and 30 Under 30 lists (Korey made the cover of the latter).[1]

From the outside, it seemed as though Korey led the Away team with a culture deeply entrenched in the ideals of Trust Totally. To reporters, Korey emphasized that she "[empowered] decision making at all levels of the organization . . . [creating] a culture that's rooted in growth and learning." Of Away's fast track to success, she said in a *Forbes* video, "If I'm going to take credit for anything, it is that I built an amazing team." From all appearances, Away had the façade, at least, of a Trust Totally workplace.

But behind the scenes, a lot of people on that amazing team had a different, darker perspective—Away had a hidden culture of shaming, bullying, and fomenting distrust, a culture led by the CEO herself. If Korey's emails to colleagues and team members or her calls to their voicemails went unanswered, her yelling was audible. *What is this shit?!* In one email, Korey referred to her team as *Millennial Twats*. The team, many of whom were often on the verge of tears, stayed silent. "We just kind of let her rant," one employee said.

Away's culture of intimidation and fear manifested in their digital culture. Anything employees wrote could become potential fodder for one of Korey's tirades or reprimands. Employees were prohibited from emailing each other, and direct messages on Slack were permitted only for small requests. Korey fired four employees when she found out they were talking about her in a #HotTopics Slack channel, a known and specific virtual discussion around business issues. There was no privacy, no place to show vulnerability without the fear of reprisal. In one characteristic message sent at 3:00 a.m., Korey informed the overworked and understaffed customer service team that they couldn't stop working or submit vacation requests until they'd resolved the customer service issues she'd identified.

Organizations that Trust Totally are a lot different. In businesses with high levels of Trust Totally, teams are encouraged to work hard, because they are **Valued Visibly** and supported as they close in on their goals. High-trust organizations **Communicate Carefully** and seldom face misunderstandings. And they **Collaborate Confidently** because they move past fear in team dynamics.

THE LEADER'S ROLE IN CREATING TRUST

In 2016, to great fanfare, Microsoft released a Twitter "bot." The company proclaimed that its new tool would help pave the way to a new era of human-to-artificial-intelligence conversation. The bot, named Tay, was engineered to be "casual and playful,"[2] but it didn't take long for Tay to be hijacked by Twitter users who, recognizing a "vulnerability" the design team had overlooked, taught the bot to tweet "wildly inappropriate and reprehensible words and images."[3] The ensuing headlines about Tay ranged from "Twitter taught Microsoft's AI chatbot to be a racist asshole in less than a day" to "It took less than 24 hours to corrupt an innocent AI chatbot." Within less than a day of its release, Tay was retired permanently, prompting Microsoft CEO Satya Nadella to release a direct, thoughtful, and humble apology, expressing "deep empathy for anyone hurt by Tay."[4]

Heads would roll—right? No. Instead of chastising his development team, Nadella wrote them an encouraging email. "Keep pushing, and know that I am with you," he said, adding that the "key is to keep learning and improving." Nadella later told *USA Today*, "It's so critical for leaders not to freak people out, but to give them air cover to solve the real problem. If people are doing things out of fear, it's hard or impossible to actually drive any innovation."[5]

Trust Totally, then, shows up through our actions and communications in times both good and bad.

WHAT DOES A TRUST TOTALLY WORK CULTURE LOOK LIKE?

- When your boss sends a calendar invite without context or texts you saying she needs to speak urgently, you don't get anxious.
- When all employees actively and thoughtfully participate in group discussions, both in person and on digital channels.

- When junior employees are willing to speak up and share differing opinions.
- When all voices are heard, yet interruptions are few during phone and video meetings.
- When bullying behavior is rare, but if it happens, it's stopped swiftly.
- When there are low levels of anxiety around digital messages because everyone respects the norms of each medium.
- When teammates go on mute during conference calls, and you don't automatically assume they are multitasking.
- When you don't receive a response in an expected time frame, you don't jump to negative conclusions.

MODEL THE BEHAVIOR YOU WANT
TO SEE IN YOUR TEAMS

What you model as a leader ultimately shows up in the culture of your teams. If you're not clear when you assign tasks and responsibilities and you later chastise your team for failing to deliver on what you wanted, you erode trust. If someone challenges your idea and you shut that person down immediately, you further erode a company-wide net of psychological safety while giving your team implicit permission to shut down other members as well.

As Scott was overseeing a restructuring at his company, he realized that he had to fire the marketing head, who just wasn't performing. However, two months later, the guy was still there. "I just can't," Scott said when anybody asked. He knew exactly what he needed to do in order to grow the company right, but if he fired the marketing head, what would happen to all the unfinished projects, and would he also risk losing all his external contacts?

Problem was, *not* firing the marketing head confused Scott's team. Was Scott *really* serious about improving the business?

A few of them began doubting their boss's commitment to making the tough calls. "Are you sure?" team members started asking Scott during meetings. Things came to a head one day when one of Scott's direct reports delayed firing one of his colleagues. If Scott was so averse to firing people, why couldn't he be too? Once Scott realized he was causing a negative domino effect, he immediately fired the marketing manager, later acknowledging in an email to his team, *I should have moved sooner, and I learned my actions matter.*

CREATE PSYCHOLOGICAL SAFETY

Psychological safety means being able to speak your mind without fearing any negative consequences to your self-image, status, or career. Without psychological safety in place in a company, no one will ever speak up. (What if they get something wrong? What if their colleagues judge them, or blame them?) Harvard professor Amy Edmondson counsels leaders to "'[m]ake explicit that there is enormous uncertainty ahead, and enormous interdependence.' [. . .] In other words, be clear that there are areas that still require explanation, and so each team member's input matters. [For example, saying things like] 'We've never been here before, we can't know what will happen, so we've got to have everybody's brains and voices in the game.'"[6] By framing your language appropriately, you set up your team as a cohesive force against whatever obstacles may lie ahead while also giving them implicit permission to speak up. Now *that's* confident collaboration.

For this to work, it's essential that leaders follow Satya Nadella's lead in addressing mistakes or bad ideas: criticize the *action* instead of the person while giving your team your unwavering support.

HOW PSYCHOLOGICALLY SAFE IS YOUR TEAM?

It's important to talk openly about psychological safety, but you should also have a process to measure it. Ask yourself, and the people on your team, how strongly they agree, somewhat agree, somewhat disagree, or strongly disagree with the statements below:

1. If I make a mistake, it's often held against me.
2. Team members are able to bring up problems and tough issues.
3. People sometimes reject others for being different.
4. It's safe to take a risk.
5. It's difficult to ask other members of this team for help.
6. No one would deliberately act in a way that undermines my efforts.
7. My unique skills and talents are acknowledged and valued.

How strongly you and those on your team agree (or disagree) with these statements, both online and off, speaks directly to the level of psychological safety present within your group. This can be an effective Trust Totally exercise for you and your team.[7]
(Modified from Amy Edmondson's Team Psychological Safety Assessment)

ALLOW YOURSELF TO BE VULNERABLE

The more emphasis a leader places on vulnerability and learning, the easier it is for team members to speak up, ask questions, and embrace the discomfort of uncertainty. Communicate simple statements—"I may be missing something—I need to hear from you"; "I'll admit that operations is not my strong suit, and I'm open to your suggestions"—that encourage your teams to speak up, and that also remind them how much you value their contributions. When someone offers feedback, accept it with grace: "Point well taken. We used to be a lot better at this and we lost sight of staff communications. I promise this will change."

During one-on-ones, try to locate places that your team may be vulnerable. Will, a team leader at Facebook, typically asks four questions of his teammates: *What are you working on? What's going well? What's not going well?* and *How can I help?* Depending on what team members need month by month, Will finds himself playing the role of therapist, coach, cheerleader, and advocate.

It can be challenging to show vulnerability, and depending on a person's title or company role, that vulnerability is often met and judged differently. For example, if the CEO asks a question, it's bound to be perceived differently than if an intern, or even a middle manager, asks the same thing. Some comments or actions are perceived differently across gender, age, and culture (we'll get into this more in part three). Nevertheless, we should communicate across these differences in whatever channel and style feels most psychologically safe.

THE TEAM MEMBER'S ROLE IN CREATING TRUST

A leader is mostly responsible for establishing a tone of psychological safety, but that doesn't mean that other team members have to feel powerless. Everyone is essential to creating a Trust Totally environment. Having said that, what can we do to increase trust on a peer-to-peer level?

Migrate from Phony to Authentic Communications

In my experience, most people don't reveal themselves all at once. Instead, they show themselves slowly, bit by bit. So how do we reveal what lies under the surface and show who we *really* are, what we *really* think, how we *really* feel—in short, what's *really* going on?

For example, you might find yourself composing a message that reads, *Hi, John, this is Robert. I was looking over your website and feel my company, Doe Corporation, has a product that fits your needs . . .* Sorry, but could that possibly sound more canned? Instead, go out there and do the work. Study the company's website, read as many blog posts as you can, and assess what the company needs. Where can you be most helpful? Also, why should they trust *you*? Once that information is under your belt, try again: *Hi John, this is Robert. First, let me tell you how much I enjoyed reading your post about the elementary school initiative this past year. What a cool way to build team spirit and give back to the community.* Go forward from there. With every scrap of detail, you begin to develop trust.

Engage in Digital Watercooler Moments

Research shows that when we transition to remote work, what we miss most are the social, relationship-building activities that take place spontaneously, like when we walk by someone's desk and say hello, converge in the breakroom to discuss our latest Netflix binge, or ask a distracted colleague if he's okay. These "watercooler interactions" are essential ingredients for building camaraderie, morale, and trust. They also keep us in the loop around what's *really* going on in an organization. So without an actual watercooler, what are you supposed to do?

The answer: create the time to just hang out and have fun together. It doesn't have to be a strictly planned social gathering; five to ten minutes at the beginning of a team meeting will do. Your team should feel comfortable acknowledging the obvious fact that they have lives outside work. One team member of an entirely remote team once told me, "Every morning we start with Zoom all-hands meetings—what did you do yesterday/what about today/do you have any blockers? We also do another at the end of the day—what worked? What didn't? What did we try? It's a great way to celebrate our successes, share challenges, and create boundaries."

At another recruiting firm, shortly after the Covid-19 pandemic sent the entire United States into quarantine, the team discussed how virtual happy hours helped members navigate the sudden shift to remote work. As one member recalled, "I had a virtual happy hour with about 60 of my colleagues. We laughed, celebrated, connected—saw some very cute kids and also pets in the background. We've committed to keeping this tradition going, which is so helpful. What a morale booster!" Elsewhere, Zoom lunches became the new social cafeteria, where employees could come together and share virtual meals. As months went on, the social Zooms generally had fewer participants, but it was good for team members to know that they were set up and open if anyone needed some social interaction.

At a nonprofit, Karen told me, "My team decided to do a 30-minute call once a week, *not* about work, but to just shoot the shit and talk. It's about having fun, talking about positive things, and hearing how we are all making use of our time."

THE TRUST TOTALLY AUDIT

Ideally, teams need some kind of barometer to assess how well they are fostering a Trust Totally culture. This digital body language question guide below can help you assess both your leadership *and* your team:

Value Visibly	• Do we feel our time is respected? • Do we feel like our best work is acknowledged and celebrated? • Do we feel comfortable voicing our concerns?
Communicate Carefully	• Do we feel there is a common understanding of priorities and next steps? • Do we have a clear understanding of which channels to use and when? • Do we have clear language and vocabulary to foster understanding?
Collaborate Confidently	• Do all the appropriate stakeholders feel identified and aligned? • Do we feel that the correct people are informed—and are they cascading the messages appropriately? • Do we feel there is consistency in communications across teams?
Trust Totally	• Do we give each other the benefit of the doubt when facing uncertainty? • Do we feel safe enough to speak up? • Do we create moments for informal social connections?
Get the full assessment at ericadhawan.com/digitalbodylanguage	

By now you might be wondering, *If I take all the necessary steps to initiate these four pillars into my team, what exactly can I expect?*

The answer? You can expect an organization that is resilient and adaptive, one that comes together in tough times as well as in good times.

When you **Value Visibly:** Team members show up at work with excitement and drive. They're *motivated* to make meaningful contributions and innovations, prompting employee engagement, retention, and productivity.

When you **Communicate Carefully:** Teams present a single, united front, get projects done quickly and efficiently, and feel safe bringing up potentially groundbreaking ideas.

When you **Collaborate Confidently:** You create organization-wide alignment on common goals without misunderstandings or petty disagreements, leading to cross-team collaboration, innovation, customer loyalty, and marketing effectiveness.

When you **Trust Totally:** You create high levels of organizational faith, where people tell the truth, keep their word, and deliver on their commitments, in turn creating client/customer sales growth and cost-effectiveness.

Here's a recap of everything we've learned in part two:

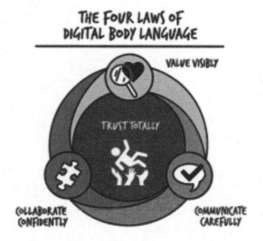

THE FOUR LAWS OF DIGITAL BODY LANGUAGE

VALUE VISIBLY

TRUST TOTALLY

COLLABORATE CONFIDENTLY

COMMUNICATE CAREFULLY

PART THREE

Digital Body Language Across Difference

Thanks to my long-standing interest in how people collaborate at work, I'm often asked to speak at events focused on business growth, teamwork, and innovation. At one of these events, a conference hosted by a large investment bank, I listened as the chief human resources officer kept emphasizing the importance of "inclusion." Then, trying to rustle up engagement, he addressed the audience, made up of around 100 newly promoted VPs, and asked:

"How many of you have ever been in a position where you didn't feel included?"

Zero hands went up. In fact, nearly everyone in the room either looked down or away. As silence washed over the ballroom, it became clear that no one planned to admit to feeling excluded in front of a white male executive and 100 of their fellow recently promoted VPs.

Consider this: *Was there a more inclusive way the CHRO could have asked this question that would have solicited honest, vulnerable responses that provided a real path forward for his company?*

Next, imagine if this same chief human resources officer asked this same question in a group email, text, or Webex. The result, unfortunately, would have been the same. Digital communication has been dubbed a great equalizer, but a lack of Trust Totally in the real world merely gets amplified online.

When we find ourselves irritated by a young colleague's too-casual tone, a delayed response from a French colleague, or a seemingly sexist reference prompted by a poorly placed emoji, it's important to ask ourselves whether we just might be misunderstanding things. Just as important is asking whether we might be sending out confusing signals *ourselves.*

What do we miss when we don't tap into our own digital teams' diversity of experiences and judgments? Short answer: *too much.*

In part three, I'll discuss how you can strengthen your team's engagement, productivity, and morale across differences with an understanding of digital body language cues.

Also, I can't help it—throughout this section, in addition to third-party research, I'll be speaking from my own perspective as an Indian-American woman and self-proclaimed "geriatric millennial" raised in the United States. Therefore, not all of what I'm about to say about digital body language differences may resonate with you. People's upbringings, personalities, power levels, and work styles inevitably add nuance and specificity to their communication habits. I ask only that you approach this section with a willingness to confront potentially uncomfortable truths and possibly your own biases. Without taking this step, organizations will always fall short of reaching their full potential.

●

CHAPTER 8

Gender

He Said, She Said, They Said

As a little girl, I watched my mother step back from a thriving career as a physician to raise my siblings and me. It wasn't always easy for her, but she was fantastic at it. I remember how challenging it was for her after we all left for college and she had no kids to take care of anymore, no one to drive to school or check on to make sure they were getting only A's on tests. I vowed to myself that someday I would do whatever it took to have a family *and* keep my professional life going—if I *could*, *that is*. I'm a twenty-first-century woman after all. I thought I could have it all—right?

Fast-forward two decades to when I found myself married and pregnant with my first child, wondering: *Wait—what if I actually* can't *have it all?* I was beginning to gain real momentum in my speaking business, and I realized to my surprise that I was reluctant to tell any of my clients that I was pregnant. What if they thought I would no longer be available after my baby was born? What if they stopped hiring me? What if my career fizzed out just like that?

So at least for a while, I concealed my pregnancy as much as I could. I relied more on digital communication than face-to-face meetings, and I cut back on speaking events. A week or so before I delivered my son, I spent 30 minutes before a Webex seminar, adjusting and tweaking every possible camera angle so I could look more *un*-pregnant. I remember feeling so grateful that a screen allowed me to continue my business.

After giving birth to my son, it was obvious to most people that I was now a mother—and my business grew unchecked. Yet I still remember how often I felt compelled to conceal my mom life behind a video screen. Reflecting back on that time, it struck me that one of the keys to building trusted relationships across genders has a lot to do with recognizing our own fears and making the decision not to project them onto others.

This isn't to say that women no longer face the hurdle of being taken seriously, because we do—especially in male-dominated industries like venture capital and technology. In 2015, when Penelope Gazin and Kate Dwyer launched an online marketplace that sold unusual art products, Witchsy, they sometimes ran into trouble communicating with customers, art buyers, and tech developers.[1] Their interactions mostly took place over email, and sometimes the correspondence they got back was condescending or even rude.

That was when Gazin and Dwyer made the decision to bring in a male co-founder, Keith. By the way, Keith didn't exist. Keith was a fictional character who was "put in charge" of all outside messaging. Unsurprisingly, and a little dismayingly, having a "male co-founder" made a significant impact on Gazin and Dwyer's business.

In fact, "[i]t was like night and day," Kate Dwyer told a reporter from *Fast Company*. "It would take me days to get a response, but Keith could not only get a response and a status update, but also be asked if he wanted anything else, or if there was anything else that Keith needed help with."[2]

It's not pretty—but so it goes.

Even before emailing and IM'ing our bosses and underlings, exchanging emojis over work email, or accidentally forwarding a dodgy email to the entire organization—workplace communication across genders was fraught. In 1990, Deborah Tannen's book *You Just Don't Understand* opened readers' eyes to the contrasting "conversational rituals" that went on between men and women. Pop psychologist John Gray kept the subject alive in 1992 when he published the lighter-hearted *Men Are from Mars, Women Are from Venus*, which confirmed, as if anyone doubted it anymore, that men and women have vastly different ways of communicating, understanding, and expressing appreciation.[3]

Today, decades later, gender divides continue to affect our digital communications. When we answer an email and assume that John Smith, who is cc'd, is the boss and Karen Barry, who actually wrote the email, is his assistant, or when we immediately reply to Tom (who makes frank, fast, matter-of-fact statements in the group email) but take 24 hours to answer Sarah (whose emails are both lengthy and thoughtful), we're showcasing our unconscious biases. Digitization has made an already tense communication landscape even *more* complicated. And let's not forget the added complexity of inclusivity. As the business world adapts to be more inclusive of people across the gender spectrum, we must consider how these changes amplify historic gender biases and give birth to further misunderstandings.

HE SAID YEAH, SHE SAID LOL!!!

While the science hasn't kept up to be inclusive of children across the gender spectrum, studies show that traditional male-female gender norms take root early on. "From age two or three, kids show patterns where little boys are more assertive and girls more indirect," says linguist Susan Herring.[4] "Already as toddlers, the idea that girls should take others' feelings and desires into consideration before speaking or acting has formed. And for boys, conflict isn't just okay, it's encouraged."[5]

As children are socialized, these gender differences, often subtly encouraged by parents and teachers, only grow. Anthropologists Daniel Maltz and Ruth Borker have found that boys and girls have conflicting ways of talking to their friends.[6] While children of all genders engage in many of the same activities, the games they favor differ, and so does their use of language. Little girls tend to play in small groups, and usually in pairs. Their social life oftentimes centers around a best friend, and close friendships usually form from sharing "secrets." Little boys, on the other hand, tend to play in larger groups, often outdoors. They spend more time *doing* things, often for the sake of status, than *talking* about things.

By the time they're ready to begin their careers, traditionally, boys have been conditioned to seek status by taking center stage, telling stories and jokes, boasting about their skills, and arguing about who is "best" and at what.[7] Girls, on the other hand, have generally been conditioned to place a premium on intimacy, and to express their preferences as suggestions rather than orders. As for girls boasting, forget about it—where's the humility in that?[8]

Years earlier, these differences played out during school recess. Now they play out in the workplace and across the digital world.

Consider the stereotype of a powerful, successful male in a corporate setting. He's Gordon Gekko in *Wall Street*. He's Don Draper in *Mad Men*. His voice is deep, his conviction absolute, his body language territorial (I'll explore "digital mansplaining" later on in this chapter). His younger male contemporaries show a marked playfulness with one another by telling jokes, fist-bumping, and playing pranks. In group settings they present their ideas boldly, and often have no qualms about stealing a colleague's idea and making it their own.

Women, on the other hand, typically seek out a smaller group of female peers of similar status (or a single best friend). More than men, they are drawn to close, tight-knit friendships. They are Donna and Rachel in *Suits*, or Joan and Peggy of *Mad Men*. They are less likely to engage in telling jokes or playing pranks. Powerful women aren't necessarily the loudest in the room, but it's almost a prerequisite that they're among the smartest and the most qualified. Oftentimes, the higher a woman's status and power, the more she's learned to adopt more traditional male body language and ways of communicating. Still, no matter how high up on the corporate ladder she sits, a woman is still more likely than a man to use team-building to get ahead (a lifelong habit).

How do traditional gender differences show up in our digital landscape? And what can we all do about them?

ACKNOWLEDGE YOUR OWN BIASES

Consider some of the gendered stereotypes at play in our socially normed world. You get an email from a woman. It's brief, to the point, and lacking any niceties. Conclusion: she's bossy, authoritative, and probably not very nice. You get a second email, but this time it's from a man. It's brief, to the point, and lacking any niceties. Conclusion: he's confident, in charge, and doesn't suffer fools gladly. In short, the exact same email can elicit different reactions based on the gender of the person who sent it.

We're not entirely to blame for our own unconscious biases. Many, in fact, are beyond our control. Harvard's Project Implicit, a nonprofit organization focused on educating the public about hidden biases, defines these as "attitudes and beliefs that people may be unwilling or unable to report," adding that, "[f]or example, you may believe that women and men should be equally associated with science, but your automatic associations could show that you (like many others) associate men with science more than you associate women with science."[9] No one, even women, is exempt from these biases, even when they contradict our own stated "beliefs." Nonetheless, understanding when and how they show up, along with their deeper causes, can go a long way toward creating better collaboration in our workplaces.

What do gender biases look like in the world of digital body language? Below is an emailed response a manager received at work:

Sandra, Loved the white paper!!! —M, xx

What's your immediate response?

Are you assuming the message was written by a man or a woman?

A woman wrote it, right? You've deciphered a range of unconscious cues—extra exclamation points to pad out emotion or a smiley face connoting friendship, the shorthand "xx" for kisses (which, by the way, is rarely used by any gender at work)—on the way to determining that the email, *no doubt*, was written by a woman.

Mind you, the above message is a perfectly okay thing to write—but many of us unconsciously read it as "female." The issues arise in additional assumptions we may make about the sender that *transcend* gender.

It's not that we don't make positive assumptions about women in the digital workplace, because we do. For example, it is often perceived that women tend to email more slowly and completely, incorporating more feedback from others, answering all previous questions, and pausing before replying. Research shows that we also believe women are more expressive than men in their online communications. Writer Elizabeth Plank said it best: "I feel like I need to be nice and warm and inviting, even if I'm having a conversation about an issue I want corrected or taken care of. When I have to ask someone to do something, I feel like I need to sugarcoat it with my lady language (!!!, emojis, GIFs) so I don't come off as shrill, bitchy, or any other stereotype about women. Men can be direct. Women often don't have that luxury."[10]

By contrast, masculine digital communication tends to be lightly strewn with short, rapid, matter-of-fact statements, often consisting simply of an "ask" or factual information. Niceties be damned! Greetings don't include a "Dear Jim," which seems overloving, and instead are tightened to "Jim" or simply initials such as "JB." Men tend to use fewer emojis and usually see no need for extra punctuation marks.

If bold, assertive, masculine communication dominates most of the online world and the world of business, competition and directness become conversational norms. Most men are expected to accept this as part of the game. At the same time, assertive women who communicate directly are sometimes perceived as cold, ruthless, or withholding.

Below I share examples of digital body language across the spectrum, from masculine to feminine styles. I recognize that these labels may feel limiting to some. You don't need to put yourself in boxes, but instead can use these examples to spot your own habits and potential biases.

MASCULINE DIGITAL BODY LANGUAGE . . .

- Confidently transfers conversations between mediums.
- Refrains from using emojis or excessive punctuation.
- Ensures messaging is short, well organized, and to the point.
- Uses bullet points and clear subject lines in emails.
- Often cc's superiors on emails, even when unnecessary, with the goal of receiving praise.
- Uses boosters to make their messages more assertive and definite (e.g., "always," "definitely," "obviously").
- Has quick response times.

FEMININE DIGITAL BODY LANGUAGE . . .

- Prefers face-to-face meetings over digital mediums.
- Uses niceties and politeness as tools, along with hedging language (e.g., "perhaps," "might," "I think maybe").
- Uses intensive adverbs, nonstandard spelling, and punctuation to reveal emotion (e.g., "sooo," "?!?!?!" "ridiculously," "Nooo waaay").
- Always proofreads messages.
- Has slower response times.

Below is a text conversation between Julie and Tiffany, two old friends who are just checking in on each other. What's going on between them?

Julie: hey

Tiffany: hi.

Julie: how are you? It's been a while! Haven't talked in forever

Tiffany: Fine

Julie: How is work? How's home?

Tiffany: Busy.

Julie: oh man, how are you holding up?

Tiffany: I'm good.

Julie: Did you get your review yet?

Tiffany: No

Julie: Alright I guess I'll talk to you later . . .

Tiffany: Bye.

Men reading this exchange had more or less the same conclusion: Tiffany was busy. Tiffany had no time for a longer conversation. (Men largely ignored any signs linked to the strength or weakness of the friendship.) Most women, however, had a different take. The lack of hedging language, the absence of detail or any back-and-forth questions—*yup, Tiffany was definitely mad at her friend*. Why? No idea. But mark their words.

We all have biases and expectations of how men and women communicate, but remember, they're not always true. They can also be subtly or not so subtly influenced by considerations, including a person's age, country, or company culture. One client told me, "I have two guys on my team. One man is very curt, and the other is the king of using exclamations. I think their styles are more influenced by their age and culture than their gender." Laine, an executive at Citigroup, added, "As a woman, I struggle to write emails to a majorly female coaching group I'm a part of. I find the group chat so effusive, it's not my natural voice," which, she explained, was more direct and concise.

Here are a few powerful actions you can take to achieve optimal clarity across genders.

KNOW YOUR AUDIENCE

If you're like most people, you've probably evolved a digital body language that feels natural to you. But it may not always work in the context of a workplace. We all need to figure out when to let our digital personalities shine and when to conform to explicit or unspoken expectations in our office cultures.

A few months ago, I was having dinner with Jake, the CMO of a Fortune 500 pharmaceutical company. As he and I were discussing the talent on his team, Jake specifically referenced one of his star players, Jessica, his 27-year-old chief of staff. Among Jessica's responsibilities was giving assignments to all team members. Jake was a big fan of Jessica's confident speaking style and project management skills in live team meetings. So why was her digital communication style so, well, *tepid?*

In emails, it seemed, Jessica had a habit of framing the jobs she assigned others as questions or suggestions. She would ask, *How would you like to do the X project with Y??* or, *I was thinking of putting you on the X project. Is that okay???* 😊😊 Jake was worried that Jessica's style and tone made her come across as immature, uncertain, lacking in confidence, or even a pushover.

In his midyear review conversation with her, Jake went so far as to advise Jessica to change her demeanor with her staff in their online communications, adding, "If you use qualifiers with the men on this team, they'll walk all over you." In response, Jessica began editing her language, using framing statements like "I'll be as specific as possible" and "To get to the point" as a way to streamline any filler language. It worked too. Tailoring the way she communicated ultimately cemented Jessica's ongoing leadership.

If you take any lessons from this story, let them be these. First, we sometimes need to adapt our natural digital body language to match the tone of our workplace. Second, a lot of our default digital body language was likely learned in past work environments and from previous relationships (meaning that Jessica will export her newfound assertiveness to her next jobs). Don't assume that someone who uses fillers or hedging language in emails lacks maturity or is in any way less credible. Instead, recognize that digital body language tendencies are likely the result of past experience—and encourage changes only if that language affects the clarity of a person's communications.

WHAT WILL HELP ME SHOW CONFIDENCE?

- Don't over-apologize. (*I'm so sorry about this; Hope I'm not bothering you; Hope you don't mind but . . .*)
- Avoid hedging language. (*Probably; I think that maybe; I guess; I'm not sure, but . . .*)
- Limit excessive flattery or subservience. (*I'm sure you are busy; Is there any way I can have a few minutes of your time?; I know you have a lot on your plate . . .*)

If you're a team member halfway up the corporate ladder, ask your colleagues what signals *you're* sending. Their answers may surprise you.

Jessica may have been told to toughen up her language, but

more often professional women are counseled to *soften* their language. The *Harvard Business Review* observed that in order to benefit from their competence, women need to be seen as warm *as well as* confident and influential. By contrast, competent men are seen as confident and influential, and it makes no difference whether they are warm, cold, or in between.[11]

Next, let's consider Sonya, a manager at a financial services firm, a position that took her over a decade to attain. Sonya has always taken pride in her communication skills—so why, at her appraisal meeting, did her boss tell her that her blunt, formal emailing style needed to be more "friendly"? Some team members, it seemed, interpreted her writing style as peremptory. Surprised, Sonya pointed out that the quantity of email she dealt with each day meant she usually had to respond quickly and tersely.

Would these same accusations be leveled against a man? Hard to say. But Sonya proceeded to adjust her language. *Do this* became *Let's try this approach. Please complete* turned into *This is what I'm thinking of as the timeline, what are your thoughts?* She added exclamation points and emojis. These changes made her messages come across as personable and collaborative rather than commanding and rude; the issue never came up for Sonya again.

Is this a sad story, or an instructive story? Maybe a little of both. The bottom line is that, in this particular work environment, Sonya could be seen as an open-minded, inclusive employee only by adopting the stylistic and tonal conventions generally associated with women.

Not surprisingly, female leaders have varying opinions about this. Some believe that if a woman has a naturally authoritative style, then she should be true to her authentic self. Others point out that a certain amount of adaptability is essential for success, even if it serves to reinforce gender stereotypes. They argue that, if a woman needs to be "nicer" and less direct to be perceived well, then she should just do so—at least until she's in a position of power where she can change things.

Speaking personally, I commend Jessica and Sonya for adapting to their situations and circumstances. Men aren't exempt from this kind of advice either. If you're a guy, take the leap and use an emoji or exclamation mark at work, especially when it might improve engagement and trust in your team. We all have the power to disrupt stereotypes, and it starts with each of us.

That said, in a viral conversation on Facebook in 2017,[12] a group of professional women shared the liberation and sheer rebelliousness they felt when they deliberately omitted signals of "female warmth."

> I always feel like I'm slow-walking away from an
> explosion when I don't use exclamations.

> My life!! I use more time deciding my punctuation
> strategy than actually writing/editing the damn
> email.

> Hahaha yessss and guilty/second-guessing feeling
> the rest of the day.

Leaders should create space for everyone on their teams—regardless of gender presentation or identity—to be themselves while also being appropriate with customers and clients. Team members are also advised to find work environments where they can be themselves and in which they can excel. As more workplaces become increasingly diverse and inclusive, chances are that women won't have to be "soft" in order to succeed. The world needs more direct women and emotional men! Or more bluntly, let's not be so sexist.

BE AUTHENTIC. BE YOU.

If you're a woman who feels the pressure to display warmth when making direct requests:

- Show both competence (direct requests) and warmth (something as simple as a friendly greeting or signature).
- Be succinct, but also offer a framing statement like *I'll be as specific as possible* to avoid a potential assertiveness backlash.
- Be direct, but also explain the motivation: *I'd really appreciate it if you'd do this. We need to get this out by 5:00 p.m. because the product is launching tomorrow.*
- End with *Best, Thank you,* or no salutation.

If you're a man stuck sending flat, toneless messages:

- Dare to use an emoji or more than one exclamation mark at work.
- Use a closing salutation like *Thanks* to end your thoughts, even via text.
- Call on women in meetings, or give shout-outs to their digital messages.

STANDARDIZE COMMUNICATION

One leader whose team is largely made up of female millennials once told me she'd banned the use of exclamation marks and emojis. She also had her team use a template of "Who/What/When" for all their emails. This eliminated the "need for niceties" that many working women feel, a need that can create gaps in a team's ability to Communicate Carefully.

Another leader I worked with established a rule of starting

emails with WINFY, short for "What I Need From You." This reduced cross-gender misunderstandings, especially among female employees who, again, felt pressured to present themselves as nice, cordial, or hedging in their requests to colleagues.

Finally, one PR executive created a norm that every work task had to be written in an email or on Slack following a phone call, eliminating any confusion that arose from varying masculine to feminine digital body language styles.

Some men have become advocates for a more direct communication style *regardless* of gender. James Fell, a blogger with a blunt, spare writing style, reports that when women who know only his online persona meet him in person, they're usually surprised. The typical response? "You're not nearly as big of an asshole as I thought you'd be."[13]

Why does Fell come across that way to his female readers? Well, he never uses exclamation points to convey excitement. (He uses them only when expressing outrage or urgency.) When working with female editors, he noticed they would pepper his text with exclamation marks (Fell always deleted them). Male editors, on the other hand, almost never did this. Fell was left with the pained understanding that women—even some female editors—often project their own "gender training" onto their work. "No one gives a shit if a man is blunt and uses a simple period to end a sentence," Fell wrote in one blog. "But a woman must use exclamation points to express enthusiasm lest she be thought a bitch and get talked to by a supervisor about her 'tone.'"[14]

STOP SELLING YOUR WORDS SHORT

Every day, the pressure on women to appear warm and friendly plays out in the form of *hedging language*—for example, "filler" words that many women embed into their texts and emails to make them appear less harsh or assertive. The most common examples include adding "I think" or "I'm wondering if" before stating an opinion, or adding "but I'm not sure" or "but what do you think?" at the end of a sentence, when in reality we're probably *completely* sure, and in fact *positive* we're right!

I'm hardly immune to this. I once had to reach out to a client about an extremely late payment. I wrote what I thought was a businesslike email and then asked my husband to read it over, as I often do with important correspondence. He kindly and swiftly ripped it apart. "You don't need to say 'just,'" he said. (After saying how I hoped she was doing well—exclamation point!—I told her I was "just" writing to inquire about the overdue payment, and "just" hoped she'd look into it.) "Say what you mean. You don't have to tiptoe around it. *She* owes *you* money, remember?" I took my husband's advice, rewrote the email, and pressed *Send*. In the end, the presence—or absence—of fillers didn't make any difference!

FILLER WORDS THAT DON'T ADD VALUE TO YOUR MESSAGES:

- It's my feeling that . . .
- I feel maybe . . .
- I guess . . .
- I'm not sure how strongly I feel about this, but . . .
- In my opinion . . .
- I guess my question is . . .
- I just . . .

Eager to be more conscious of my own hedging language—and maybe someday even eliminate it—I stumbled onto a Gmail plug-in called *Just Not Sorry*, which uses an editor's red strikethrough line to highlight hedging language in any email. A steely, tough-minded reality check, *Just Not Sorry* lets users hover over an underlined phrase and explains how others are likely to perceive it. Well, *Just Not Sorry* underlined many more sentences than I expected—and taught me what years of business experience couldn't. Today, without even thinking about it, my emails are (mostly) clear and direct.

DESIGN SPACES FOR UNHEARD VOICES

Creating psychological safety across the gender spectrum begins with leaders who define a company's communication norms, from medium selection to meeting etiquette. Allowing your team members to have a say in the communication channels where they feel most comfortable encourages Trust Totally.

As a manager, if you don't know already, you have to understand that people oftentimes need different things in order to feel Valued Visibly and be able to Communicate Carefully. Phone call? Email? Private or group message? A text? A personal meeting? We might not always realize it, but the choice we make isn't just based on whether or not it's practical, or fast. Our decision also conveys the level of interest we have in hearing from different voices on our team. As everyone knows, some team members are eager to speak up in a face-to-face meeting, whereas others hang back, preferring to express themselves more comfortably in a virtual chat room.

By closely studying these divisions, you'll discover that

many are gendered. Below are a few practical ways to implement structures that can amplify unheard voices:

GATHER INPUT FROM MULTIPLE CHANNELS

Communicating online can create advantages in overcoming the perceived (and often real) biases against women that make many of us feel too inhibited to say something. Research studies from Carnegie Mellon show that female students are more likely to ask their professors questions online than in their offices.[15] Similarly, in my experience, women in lower parts of the corporate hierarchy are often more likely to provide input by email than speak up in face-to-face business meetings.

As linguist Naomi Baron points out, digital communication serves an important function for women. "Using written communication online allows you to take on any persona you want, with anyone. You can mask your gender, you can mask your accent or dialect."[16] Digital communication also allows women to avoid uncomfortable confrontations, Baron adds. Why? Because text-based communication de-emphasizes traditional signs of confidence and leadership, such as the timbre of a voice. In male-dominated workplaces, it offers an effective and even unprecedented way for women to share power and decision-making authority—a phenomenon that can begin to level the gender-imbalanced playing field.

Still, the issues of youth, and even femininity, remain. I'm no stranger to either. Ten years ago, when I started my business, I found myself in my early thirties, a player in the thought leadership space that was otherwise composed mostly of older white men. I began noticing that when I used either the phone or email to sell my services to senior executives, I got twice as much business as I did when selling my services in face-to-face meetings.

Counterintuitive? Sure. Isn't trust built up in person? Later I concluded that my youthful face and not-always-gregarious body language brought out the bias of older executives. During face-to-face meetings, I heard more questions like "How long have you been doing this?" or "Can you send me more client reference names?" I got the doubtful questions less often on the phone. Even in video calls, my onscreen presence felt ageless and my ideas mattered more than my visual cues. We discussed the content I was planning on delivering and the value I could offer, and what's more, I always negotiated better fees. Today, I *still* do face-to-face meetings, especially when I haven't met the person yet, but experience has taught me that calling or emailing someone to follow up defuses potential gender bias, or outright sexism, since prospective clients lack the distraction of visual cues.

In general, as I said earlier, women create tight-knit, conversational communities online and off. Just as often, they use direct messages to vent, (digitally) roll their eyes, or make fun of male-gendered communication. *Quartz* reporter Leah Fessler once wrote, "The pressure that exists in public channels to preface or cushion comments and links disappears in DMs and private groups."[17] Fessler added, "I, too, become far more direct, and simply myself, over DMs."[18] While it's true that many mixed-gendered groups foster support and encouragement, the fact remains that many women still feel more at ease conversing among, well, *other* women. Fessler also noted the differences in positive reinforcement that she's observed when women run editorial discussions: "Not once does a woman shoot down her colleague's idea, force her opinion on the group, or ignore the comment that came before. We asked questions ('is there a way to have made it clear that it was inappropriate without getting burned') and invited in colleagues with relevant knowledge ('that sounds like something @aimee would have thoughts about')."[19]

Before choosing a primary mode of communication, leaders should consider asking everyone on their team which mediums they prefer—or simply present different options. Use a poll or survey to gather preliminary thoughts from team members before meeting. *What's the best platform to discuss this? Phone conference, Zoom, or something else?* If you're not a team leader yet, take the initiative by asking your boss about his or her preference. One deputy commissioner of an organization once told me, "The first time I meet a new boss, I ask about how to best communicate with them. Give me your instruction manual. Be so effective in your communication based on their style and get into the position where you can create the culture."

PASS THE MEGAPHONE

An executive I know once carried out a month-long experiment designed to track how often women spoke up in his meetings versus men. He was shocked by the discrepancy between the number of times in total men spoke up first compared to women. The men consistently spoke up first more often. From that point on, he made a concerted effort to call on women in meetings to broaden the perspective of the entire company.

Blogger and entrepreneur Anil Dash tried something else. Using Twitter analytics to determine the gender breakdown of his followers (who number over a million), Dash discovered that he followed a roughly equal percentage of women and men—but retweeted men *three* times as often as he did women.

So he tried an experiment.

For a whole year, Dash made it a point to amplify the broadest possible variety of voices he could by exclusively retweeting the comments of women—and recommended that others try the same experiment.[20] Said Dash, "If you're inclined, try being mindful of whose voices you share, amplify, validate and promote to others . . . we spend so very much of our time on these social networks, and there's so much we can do to right the wrongs we've seen in other media, through simple, small actions."[21]

As for any lessons he might have gleaned? "More broadly, I found the only times I even had to think about it were very male-dominated conversations like the dialogue around an Apple gadget event. Even there, I'd always find women saying the same (or better!) things about the moment whose voices I could amplify instead of the usual suspects."[22] Dash also took note of the inclusivity of his discussions: "One thing that has happened is that I've been in far more conversations with women, and especially with women of color, on Twitter in the past year."[23]

When I re-created this experiment myself, I realized I have the opposite problem—on social media, I amplify many more women than I do men. It turned out I could be a *lot* more inclusive too, and today I'm a lot more conscious about equalizing an array of voices across gender, generation, and culture.

Think about your own practices at work. Whether you're drafting an email message or preparing a team call, carefully consider the people whose voices you could potentially amplify. What unconscious assumptions might you be making about your audience? Remember, though, it's what we *do* with the answers to these questions that increases the levels of clarity and understanding in our workplaces.

USE INCLUSIVE LANGUAGE AND IMAGERY

Hey guys

Inclusive language matters more than we think. I use this phrase a lot in phone calls, emails, texts, and live meetings. I don't even think about it. Maybe that's the problem. But over the course of doing research for this book, I realized I was unintentionally excluding the women on the team. What's up with *that*?

I'm not alone. Many of today's most commonplace idioms tend to be more masculine, at least in the workplace.[24] At Buffer, a well-known technology company, the team has become more mindful of the language they use and focus on how to make it more inclusive. As Buffer was growing and ramping up their hiring, leadership there observed a very low percentage of female candidates applying for developer jobs—less than 2 percent of all candidates.[25] In response, they realized they needed to fix their job descriptions. They identified and called out exclusionary words like "hacker" in their job descriptions, with CTO Sunil Sadasivan even broaching a discussion with the team about finding another word entirely.

The solutions they came up with ranged from terms like *creative connector*, *engineer*, *developer*, and *product designer* to *maker*, *artisan*, *architect*, and *code experimenter*. Buffer concluded that *engineer* sounded the most neutral, whereas *developer* was the friendliest, clearest, and most inclusive. A shift from "We are a dominant engineering firm that boasts many leading clients" morphed into "We are a community of engineers with many satisfied clients." Qualifications changed from "ability to perform individually in a competitive environment" to "collaborates well in a team environment."[26] For Buffer it was *that* simple.

Then there's Textio, a company that uses customers' hiring data to help figure out if a company's language is gendered. Textio, for example, has identified "work hard, play hard" as masculine and "we value learning" as feminine.[27] Words like *enforcement* and *exhaustive* skew masculine, while *transparent* and *catalyst* skew feminine.[28]

Companies that use Textio to craft language that's more inclusive find that 23 percent more women apply for jobs, and 25 percent of applicants are also more qualified than the previous job candidates who were drawn toward the original role description.[29]

Being inclusive also means being mindful of stereotypical imagery—and resisting it. When the white shoe law firm Paul, Weiss posted a photo on LinkedIn in 2019 of its newly minted partner class, it set off a storm of controversy and public debate, even inspiring a *New York Times* article.[30]

Why? Because, of the dozen individuals in the photo, eleven of them were white and male, with only a single woman appearing near the bottom corner. Today, as public reaction showed, homogeneity like this is unacceptable. Nearly 200 general counsels and chief legal officers from companies including Toshiba, NEC, and Heineken signed their names to an open letter on Twitter, calling on Paul, Weiss and other law firms to step up to the inclusion challenge or risk losing business.[31] To their credit, Paul, Weiss issued an apology and shared next-step actions to incorporate more diversity among their partners.

Bear in mind that PowerPoint slide images, photos of the leadership team on a company website, and even colors chosen for a visual presentation can affect our perception of a business as exclusive—or inclusive.

COMBAT DIGITAL MANSPLAINING

Many women are raised and conditioned to build consensus. Many men, on the other hand, *aren't*. Even when they lack expertise, men are encouraged to speak authoritatively—which also extends into their online behavior. Australian feminist and author Dale Spender calls the often-patronizing ways in which men try to "explain" something to women (who may actually know more about the subject than they do) "digital mansplaining."[32] Many men are simply accustomed to grabbing more airtime in conversations, and if a woman is present, either they'll interrupt her or talk right over her.

In a digital workplace, this behavior only gets amplified. In a viral *Quartz* article by Leah Fessler called "Your Company's Slack is Probably Sexist," she noted that men are more likely to declare their opinions as facts and send along a link to an article without comment or sometimes even context. Women by contrast typically explain why they are passing along a link—*per our previous conversation about climate change*—or another way to explain why the link might be of interest to the recipient. Says one female Slack user about the comments of her male co-workers, "They just toss [a link] in because their interest in it was enough to warrant sharing it—they're assuming you'll receive their gift with graciousness, then they walk away."[33]

I belong to a Facebook group of peers in the professional speaking industry. It's evenly divided between men and women who come together every so often to share advice. One guy in our group, whom I'll call Dan, never responds to questions, but often posts his opinions. He's not there, it seems, to engage or help. He simply wants an appreciative audience. We all know to avoid him and ignore his behavior. From Dan and others, I've observed that digital mansplaining isn't *just* about interrupting people, it's also about a person conveying an un-assailable entitlement in tone and style.

MANSPLAINING IN A DIGITAL CONTEXT

A digital mansplainer will . . .

- Ignore an email from a colleague and bring it up later as his own idea.
- Deliver group work to a superior without cc'ing or crediting the team; use "I" instead of "we" when summarizing accomplishments.
- Use masked and condescending language with peers in email (e.g., Good work or Wow, not bad!), implying he's in a leadership role for which he may not be qualified.
- Jump into group discussions without background, shooting down a colleague's idea, forcing his own idea, and/or ignoring previous comments or questions.
- Join conference calls or group message conversations late while jumping in as if already fully informed.

According to researcher and linguist Susan Herring, the male tendency to mansplain is timeless—and could be seen at the dawn of the internet age. In the early 1990s, for example, Herring joined a listserv comprising more than 1,000 other linguists. "Many were claiming that online, gender and other social differences would be invisible; you wouldn't be able to tell who was who, or judge anyone based on their identity," she recalls.[34] That wasn't the case though. The online discussions that Herring followed tended to be divisive. One especially caught her eye, as the topic in question had broad appeal across the entire linguistic community and historically prompted numerous, valid opinions from both genders. "However, it was almost entirely men engaging," Herring remembers.[35]

Wondering why the women in the group were holding back, Herring sent out a survey. When the results came in, nearly all the female respondents reported that they disliked the contentious style and tone of the digital discussion, and found participating in it to be unproductive. Herring finds this same dynamic on crowdsourced Wikipedia articles,[36] concluding what a lot of people (especially women) know already, namely that "some contributors, anonymous and otherwise, use rude and haranguing language. Such environments are—if not outright intimidating—unappealing to many women."[37]

That said, is there such a thing as "womansplaining"? Does it go both ways?

In my case, at least, I admit it might. Every year, for example, my family plans a group vacation. A few years ago, during a particularly busy time in my career, my husband volunteered to take over my traditional duties as chief vacation officer. Reluctantly I agreed, assuming he would be terrible at it.

Over the next few weeks, in addition to juggling my work schedule, I insisted on reviewing every vacation option he came up with. Had he checked to see if the hotel offered free breakfast? Were there photos of the rooms he'd reserved? *Hold on a second—was I mansplaining, no, wait, womansplaining?* I was. I was interrupting and talking over his plans and ideas with my own plans and ideas, which I secretly suspected were superior. Even when he and I agreed on something, I insisted on saying it louder. Yeah, I can be a micromanager and a know-it-all, and aren't those secondary characteristics of a man- or woman-splainer?

Is there any way to shut a peoplesplainer down? Yes. Managers can stop digital interrupters from hijacking a phone or video call by being firm about who speaks, and for how long. "Follow good chairing protocols," advises André Spicer, an expert in organizational behavior at Cass Business School in London. "At the beginning, say: 'This is the purpose of the meeting, this is how long we've got, we're going to spend this much time on each item, and here's how we'd like you to share.'"[38] Simply being more aware of who is "loudest" in messages, on phone calls, and in meetings will help you guide your team to **Collaborate Confidently** by making sure everyone gets sufficient airtime.

The digital workplace flattens many traditional gender biases we've known for years. Women can be more resolute, and men can realize there is new space to show warmth and affection. At the same time, certain traditional gender norms are amplified, such as women who still feel the need to be "liked" by peppering their digital communications with exclamation marks and modifiers. Perhaps the greatest advantage is that our digital body language provides an accurate visual mirror that reflects what has occurred for so long in spoken cross-gender communication. Maybe, looking in that mirror, we can ask ourselves: How can I just be myself?

CHAPTER 9

●

Generation

Old School, New School

A female client (who is ten years older than me) once emailed me, *Talk in 30?*

I replied, *I can talk now, or in 2 hours!*

The response that followed—an extremely terse-seeming 2. (and yes, that was a period at the end)—shook me.

Was I about to lose valuable business? But our conversation couldn't have been better, and our business relationship continued.

It turned out she was just being professional. Who was I to interpret things any differently? The answer: *I was younger.*

Different generations don't merely use different digital body language; they also have different interpretations of the *same* digital body language cues. A 30-year-old woman is likely to perceive the same text message differently than a man 30 years her senior. One generation's expressions of joy, or considerateness, can be another's expressions of immaturity, or rudeness.

Typically, these divides stem from a lack of familiarity with each other's specific digital body language signals and cues.

Digital natives came of age learning the conventions of digital body language, often assuming that the signals and cues all around them were, and still are, obvious to most people. They're not! Digital adapters have had to learn digital body language as adults. For many, it can be as hard as learning a second language.

Digital natives can be seen by their adapter counterparts "as technologically sophisticated multi-taskers capable of making significant contribution, but with a communication deficiency."[1] The "deficiency" part comes from their reliance on remote work and informal, technology-dependent modes of communication that often leave digital natives unable to interpret physical body language. But digital adapters have their own communication deficiency—they're just not as good at tech![2] That said, the division between digital adapters and natives isn't always based on age alone. I've met 28-year-old digital adapters who insist on talking face-to-face about everything, and I've met 50-year-old digital natives who will respond to emails and voicemails with texts.

A client of mine once complained about her digital native sales rep. During meetings, he just couldn't read clients' body language cues. He seemed blind to their every posture and gesture, had bad eye contact, and overlooked any number of micro facial expressions that would have told him he was going seriously off track to the point of losing the client. He commonly used "so" to begin thoughts and sentences, an almost real-life mirror of ongoing, never-ending text threads instead of professional spoken conversation.

There are other stories too—like the one I heard about how

tough it was for junior bank tellers at a large financial services firm to master the range of questions customers asked them. They weren't lazy or entitled—they simply had no idea how to use the "new" tech they'd been handed. Most of them had never *not* used a mobile phone. A landline—what was that? They were clueless about how to talk to strangers. Should they put their customers on hold? What if the person got angry? When their manager finally figured out why the team seemed so confused, she retrained them on customer service etiquette, and things went more smoothly.

Growing up, I was taught how to answer the phone politely, and if the call wasn't for me, to take a message. It wasn't until I hired people who were born after 1990 that I realized I was among the last generation to learn that skill. Sam, one of my new hires, for example, had no idea how to take down messages:

> Sam: Someone called.
>
> Me: Who?
>
> Sam: Bob.
>
> Me: Bob from Idaho? From Minnesota?
>
> Sam: Not sure which . . .
>
> Me: What did he say?
>
> Sam: He asked you to call him back.

I had to email both Bobs to figure out which one had called. Becoming fluent in different communication styles is a key skill for today's leaders. "IM and texting do not come naturally to me," noted one digital adapter, an executive in a technology company. "But it's the way so many of my younger colleagues communicate, at every level of the company. When they say, 'We'll talk later,' they usually mean, 'We'll IM later.' They often have three conversations going at once. I fear that it makes our conversations more shallow, but at the same time, I have to adapt and meet them where they are."

TAKE A WALK OUTSIDE YOUR COMFORT ZONE

Good leadership is more than just about bending people to your standards or norms; it also involves a willingness to engage across the different digital body language styles present in your workplace. It's really no different from knowing three or four different languages or regional dialects.

Brad, the SVP at a large gaming company and a digital adapter, has observed a stark difference in the two Slack channels run by his leaders, Allie and Dave. Dave, a digital native, has a Slack channel filled with emojis, GIFs, and memes, whereas Allie, who is a mid-forties digital adapter, has a more formal writing style, complete with bullet points. "With Allie's Slack channel," Brad says, "I'm at home." Nonetheless, he soon came around to the way Dave saw the world. "He is so authentic. If I was to force him to be 'corporate,' his team would be less excited and engaged." He adds, "I've learned that the best thing for me to do is try to become conversant in this 'dialect,' even if it's uncomfortable."

It's a smart decision. Pause for a second before you decide to adjust how someone on your team is communicating, and consider how that person's style might end up benefiting your teams.

YOU DON'T KNOW WHAT YOU DON'T KNOW— ASK FOR HELP

Bob McCann, a management communications professor at the University of California, Los Angeles, points out that the proliferation of new technology has accelerated the growth and depth of today's generational divides. "Every three weeks, we have a new platform that we need to deal with, a new app that's coming out, and we have to adjust, and we have to change."[3] On a practical level, this means that there will *always* be new technologies around the corner—including new ways of greeting one another.

Take the subtle differences in an email greeting—*Hi*, *Hey*, and *Hello*. For digital natives, *Hi* and *Hello* are basic professional greetings, but when digital natives are among peers, *Hi* and *Hello* feel slightly formal (is the other person upset with us?). More informal and popular is *Hey*, which starts most conversations and conveys friendliness and camaraderie. Similarly, *okay* comes across as normal and friendly, whereas *ok* just might convey frustration or anger.

Shorthand like *LMK* (*let me know*) and *TL;DR* (*too long; didn't read*) is ubiquitous among digital natives. For digital adapters, it can be bewildering. Is your organization made up mostly of digital natives, or digital adapters? Do you have standardized norms of communication? What planet are you on?

Here are the most common digital body language differences between digital natives and digital adapters:

HOW DO I KNOW IF I'M A DIGITAL NATIVE OR DIGITAL ADAPTER?

YOU'RE LIKELY A DIGITAL NATIVE IF YOU PREFER . . .

- Texting back and forth even when setting up a call or meeting might be easier.
- Texting to ask if you can call (instead of just calling). Texting to tell someone that an email has been sent, rather than just waiting for a reply or cc'ing.

- Responding to a phone call with a text or email instead of calling back.
- Leaving voicemails unread and unanswered.
- Avoiding phone or face-to-face meetings in general.
- Being more responsive to social media posts than direct email requests.
- Using shorthand like *LOL Thx ttyl* (talk to you later) or *kk*.

YOU'RE LIKELY A DIGITAL ADAPTER IF YOU PREFER . . .

- Insisting on a call or meeting in lieu of a text or email.
- Not answering texts quickly (e.g., within an hour).
- Asking for details from an email to be summarized again, this time verbally.
- Using formal language and punctuation, including "signing off" at the end of a text as if it were an email or letter.
- Sending overlong emails without hyperlinks or relevant information.
- Sending curt text messages that lack context and seem alarming to natives—*I'm worried. Call me.*

EXPERIMENT WITH CHANNELS ACROSS GENERATIONS

Another generational challenge? Medium preferences. Just as different genders prefer different channels, different generations do too. Sure, some digital adapters are willing to work with new mediums, but it's usually to graft old conventions atop them. My dad, for example, sends me lengthy texts, beginning with *Dear Erica,* [*enter long letter that you have to scroll down to read*] and ending with *Love, Dad.* These texts usually arrive when I'm at work and can't write back. I've learned to respond, *Thanks for the text dad- will call later.* It's a little silly, but I like connecting with him this way, and it brings us even closer. (I've never sat him down to explain that texting isn't the same as writing a letter, and I probably never will.)

Even basic phone calls can be loaded. For example, digital adapters seldom treat a ringing phone as an intrusion. Digital natives, in contrast, do *not* appreciate getting a call out of the blue. It feels presumptuous and potentially alarmist. The way they see it, callers should first request permission to call via text or email, or else schedule a call in advance using a calendar invite request. When a digital native gives you their cell phone number, they are providing implicit permission for you to text them. For natives, texting with a new acquaintance is less invasive than getting a phone call out of the blue. For digital adapters, texting can feel invasive—like it's stepping across the boundaries of an "intimacy firewall."

Take the experience of Dana Brownlee, a corporate trainer, as recounted in *Forbes* magazine. In one of her workshops, an outspoken 50-something woman sounded off over the issues she faced in communicating with her team, which comprised all age groups. The younger people never answered phone calls and instead responded using text messages or email. *Forbes* noted that the 50-something woman "worked herself into such a frenzy that she suddenly spouted, 'We need to stop emailing and pick up the %^$# phone!'"[4]

What if the shoe is on the other foot? Younger generations share a similar frustration toward the technology favored by older generations. To them, it's not only antiquated, it also gets in the way of good working relationships. "I would never hire someone who still uses a Hotmail or Earthlink address on their resume," said Brian, a manager in his early thirties. "It just tells me they are completely out of date." While older generations may see younger employee behaviors as entitled, the younger generation sees theirs as "old fashioned" and not as productive to the current times.

But whether you are a digital adapter or a digital native, knowing the communication channel preferences of your customers or client audience is critical. Adette, the CEO of an experiential design company, once hired a sales coach to grow her company. In his late fifties, the coach was a digital adapter who kept pushing Adette's team "to hit the phones and pester your prospects for meetings. Leave a voicemail if they don't answer." Adette remained skeptical, especially since she knew that her clients (most of whom were in their thirties) mostly texted her, and would in all likelihood ignore phone calls. Her instincts were right. "Not one person picked up the phone or responded. It failed miserably. So we trusted our original intuition and decided to ask for permission before calling. We emailed people and said, *Hey I've been trying to reach you, left a vm but who listens to voicemail, right? I'd love to share our new offering with you.*" Then, instead of composing an open-ended email wondering about availability, Adette used Calendly, a calendar program that showed all available times, skipping the tedious back-and-forth scheduling. As far as booking sales meetings was concerned, it seemed that the strategy with the *least* human interaction delivered the most success.

Since then, Adette has developed new rules of engagement for her multigenerational teams. "If you need me, text me, and we can get on a call. If you leave me a voicemail, I will never hear it. I tell them I'm better with email because it's a better reminder for me to respond, versus text which can get lost when I'm in transit running between meetings. If it's urgent, ping me on Slack and tell me you emailed. If it's anything that requires complex thinking, it needs to be in an email."

Adette has also created code-switching rules for her team

to use when more formality is called for—the digital equivalent of "overdressing," as she puts it.

"A lot of our clients, who are older than us, conduct business over texts, and we hate it because there is no paper trail and they are ephemeral. If we get approval by text that has a budget typo, that's a $50,000 mistake. Our protocol is for the receiver to take a screenshot of the text from the client and put it in an email and write, *hey taking the convo to email so we are all on the same page, and take it from there.*" Adette's protocol also anticipates that employees will cc the right people while always ensuring that someone can cover for them.

Colleagues can also become frustrated with how their peers or superiors share information within the same organization. One digital adapter, Sylvie, complained about her 30-year-old co-worker: "He frequently bypasses the chain of command in emails and cc's our boss instead of going through proper channels. He'll set up time to sit with me and get information, then up and leave right in the middle of our conversation if he gets an IM that answers his questions. It's like he's got what he needed and doesn't see a need to be there anymore. What he doesn't see is how insulted I feel."

Some digital adapters mourn the loss of in-person meetings, which can build up camaraderie and create opportunities for mentoring. As one of my clients told me, "Once upon a time, junior people would come talk to me rather than shoot me an instant message. I remember the days when if someone started a discussion, they finished the discussion."

But many digital natives just don't have the patience or the attention span for in-person meetings. One told me, "Every time I ask my boss something with specific questions, I get long, drawn-out answers. I don't understand why he can't send me a quick email update so I can get back to work."

The rifts occasionally created by intergenerational digital miscommunication have wider implications too. Stress. Loss of morale. Frustration, leading to disengagement. Losses in productivity, innovation, and a company-wide sense of belonging.

What to do? First, ask yourself: What's the risk in letting digital natives or digital adapters *be true* to their own styles? If it affects the bottom line or negatively affects customer perceptions, it may be best to opt for creating explicit norms across all generations. But if, in fact, it enlivens or revitalizes your team without causing any business harm, why not step outside your comfort zone?

. . . .

Effective digital body language is about tailoring communication—not to fit the natural preferences of one generation over the other but to meet the demands of the task at hand.

. . . .

A key difference between native and adapter digital body language styles centers on formality. A study conducted by Grammarly "found workers under 35 were 50 percent more likely than older workers to be told their tone was too informal, even though more younger workers said they spent time agonizing over meaning, tone and grammar in their emails."[5]

Why is this? First, a surprising digital rule of thumb is that whenever the latest, most informal channel of communication comes along (e.g., texting), the channel preceding it (e.g., email) becomes obsolete overnight—at least in the minds of digital natives—formal, inefficient, and a potential source of fear. Younger people typically regard email as a formal mode of communication, which is why they often append sentences like *Hope your weekend went well.* Their greetings might also include a *Dear Mr. Ettling*, and their sign-offs skew toward *Sincerely*. So when older people respond with friendly one-liners, digital natives are often put off. One Gen-Xer told me that she once used the word "adorbs" in an email to a younger colleague, who responded, *That's way out of my comfort zone in email*.

For older leaders, email is an informal means of communication, especially when communicating with younger colleagues. Notes one executive at Citigroup, "I'm not naturally flowery in email, but I use many more smiley faces than I ever thought I would use at work. With my junior team, I will keep the 'two hugs' and a *Hey, how are you?* but with someone more senior I won't worry about it."

As always, paying attention, and learning as you go, can go a long way toward bridging the generational digital body language divide. Tricia, a head of HR at a technology company, told me about the varying levels of formality, views of organizational hierarchy, and assertiveness among digital natives and digital adapters. "As a Gen-Xer, there are many elements that I learned about myself from working with millennials. For example, hierarchy and relationships. For me, if there were senior-level meetings one or two levels above, earlier in my career, I would wait for the invite. In the last five years, the Gen-Yers, just ask me if they can attend meetings I am in with senior leaders. They don't wait for permission or an invite. For me, that was at first an annoyance and, later, a learning! That exposure would prove to be valuable for them in their work and their careers. I had to understand my own historical learning about hierarchy and how to shift it, for myself and those on my team."

Tricia went on to share how the video meeting culture varies as well. During the switch to remote work in 2020, many employees at her company showed up quite casual on camera without being conscious of their backgrounds. "One person even did a client video with a messy background that I believed was not professional. It was a moment where I had to understand if it was my bias or not." In that example, she gave feedback. She now reminds her teammates to fix their backgrounds on external or customer calls, but on internal calls, they can be a bit more lax. "We have to be willing to understand the new formats and how people learn to adapt. And also recognize when to discuss standards. Maybe my view on standards is a generational thing or an HR thing." Probably both!

That also extends to digital adapters feeling comfortable using newer technologies like Zoom. Many employees at her company were hesitant at first to join video on Zoom, she tells me. "I am not sure if those hesitant were natives not seeing the need for video and adapters believing it was important, or the other way around? Eventually maybe half of the company's employees adopted the platform. Once that tipping point was met, video chats became normal and less intimidating. It was an equalizing moment for both digital natives and digital adapters."

Tricia also remembers the moment when she started using emojis in emails. To her, it felt like a moment of shifting communication styles, aligning with the digital natives. Eventually emojis became the norm in her emails: "I remember the time I actually deleted the exclamations in my email and moved to the emoji."

Adette, CEO of an experiential design company, finds that familiarity works great if your team is made up of digital natives. But it's not without challenges, especially when people don't know when to become more formal. "It's a slippery slope because my employees address the senior partners in a way that is familiar, but we have to follow protocol. For example, you can't request a day off in a text; we have a process for that. A recent new hire—taking cues from more tenured members of the team—addressed one of her first emails to me as 'hey fam,' and I was caught off guard since she'd been on the team for only a couple of weeks. One time, a team member told me I was 'really formal' when I gave a Keynote presentation in a meeting instead of having one-on-one conversations." When it became clear that her team had no idea where the lines were, much less whether or not they were crossing them, Adette made her expectations clear.

EMBRACE THE EMOJI REVOLUTION

For digital natives, emojis are more than just extraneous flourishes—they comprise their own *language*, one that conveys genuine human emotion. My advice for teams? Leverage the power of emojis. Not only have they lost their aura of frivolity, they also make messaging more efficient by conveying the intent and context that may otherwise be absent.

At the enterprise cloud software company CircleCI, emojis, in fact, have become company policy. Slack posts are categorized by an emoji at the start—for example, a teddy bear emoji kicks off the meeting minutes, and team wins are marked by a thumbs-up.[6] These clear-cut visual indicators make it a lot easier for team members to find information pertinent to their own work.

Today, about one-third of young professionals feel no qualms about using emojis when communicating with colleagues, direct managers, and even senior executives.[7] And more than 60 percent of individuals aged 35 and older self-identify as frequent emoji users.[8] So, don't be surprised when your mother or even your grandmother adds a heart-eyed emoji to her Facebook comment.

Nevertheless, there are still resisters. One baby boomer client of mine confesses that when emojis show up in emails from younger employees, his first thought is, "You can't even write a full sentence. You must not be detail-oriented." I told him to let it go. Emojis may not be appropriate in any client-facing communications, but in almost all other instances, they're here to stay.

In fact, Virgin Hotels once carried out a study to determine why their newer employees weren't engaging on their internal corporate newsfeed. It turned out that some digital natives preferred symbols over words, like sending a thumbs-up emoji instead of saying "I like this initiative."[9] Virgin proceeded to do a smart thing. Not only did they embed push alerts to notify employees of upcoming events, they also increased color contrast and imagery to draw attention. In no time, engagement among new employees shot up.

The key to good communications across generations? Understanding preferences, and knowing when to adapt to others' needs and when to set proper boundaries. I promise: simply having a candid conversation about different communication styles will make a huge difference.

CHAPTER 10

•

Culture

Lost in Translation

I'll never forget when my family met my husband's family for the first time. Rahul and I were dating but not yet engaged. At the time, understandably, I was a little nervous, having met individual members of his family, but never as a unit. Still, I assumed, and why wouldn't I, that everyone would get along great, since weren't our families more or less alike? Mine came from the Punjab state in India, and Rahul's family was from India's Uttar Pradesh. We were both Indian, right? What could go wrong?

That night, we all met up at a restaurant and the introductions were made. The mood was jovial, welcoming, and relaxed, the food good, the conversation animated. I *did* notice that Rahul's parents seemed a little formal, but overall I thought the evening went well.

"So, how do you think that went?" I asked Rahul later that night when we were alone.

He paused. "How did you think it went?"

What a strange thing to say. It put me on alert. "I thought it went pretty well," I said.

"I know, because your family asked to split the bill . . ."

"What?" I asked, somewhat surprised by the sarcasm in his voice.

"In my family," Rahul said, "we don't split the bill."

"Oh," I replied, feeling suddenly awkward. "Because in my family, splitting the bill is common at a first meal."

My father has always been very generous—but usually he manages a first meeting by dividing up the bill. Rahul's more traditional family found this gesture disrespectful. Today my husband and I laugh when we remember that night, but at the time it taught me a lesson, namely that even when cultural divisions are small, people will always communicate differently.

Consider these scenarios:

- After moving from Germany to China for work, Nora was prepared for culture shock in her day-to-day interactions, but not this one: at work, instead of asking for her email address, her new colleagues requested her instant messaging account information. If and when Nora *did* get an email from one of them, she was surprised by their overall chattiness, smiley faces, and friendly greetings versus the Germanic, to-the-point style she'd grown accustomed to.
- Working with a team in Brazil, Sam, who was from the United Kingdom, assumed it was basic courtesy to preface his opinions on work deliverables in emails with words like *Unfortunately this is* . . . or *I regret* . . . His more informal Brazilian colleagues, however, found Sam's language off-putting.

- John, based in northern California, sent his peer Arvind (headquartered in India) a work request. He later found out that Arvind's boss, Raj, was furious that John hadn't checked in with him first—which puzzled John, until he discovered that Indian custom was to first ask the boss if he thought his report could spare the time to complete a request.

· · · · · ·

We don't even realize how much the culture we live in and the stories from our upbringing influence our communication styles. Did we grow up with English-speaking parents? What were the social or cultural norms of our classmates, or of our community? We also adapt signals we pick up from the culture around us. Put them all together, and they combine to create our "natural communication style." When someone else's communication style falls outside of what we're used to—too loud! too quiet! too stuffy! too punk rock!—we often pass negative judgment on them, sometimes without stopping to consider *why*.

Phan, for example, grew up in Cambodia and immigrated to the United States at the age of 12: "For me, because I'm an immigrant, I had to consciously learn how to speak like an American," she told me. The challenges she faced in written and spoken English were only amplified in digital communication. "If you can't translate a thought immediately in your head and quickly say it, you will lose your audience on phone and video. Then the awkwardness of tech difficulties, and the slight lag time, makes it even worse."

As we found earlier with the eggplant signifying one thing in one culture and an entirely different thing in another, even emoji meanings vary across cultures. The "happy" emoji, a signal of joy in America, confuses many Japanese people, who equate frequent smiling with low intelligence.[1] In China, the waving hand emoji has nothing to do with a greeting; instead it signals you're "waving goodbye" to a relationship.[2] In Middle Eastern countries, the emoji of conjoined palms is a religious symbol, whereas in Japan it simply means "Thank you,"[3] and in the United States it's often used to symbolize a high five.[4]

THEY GO HIGH, YOU GO LOW

Experts who study cross-cultural communications generally divide the world into "high-context" and "low-context" cultures. *High-context* cultures are those that communicate in ways that are implicit and rely heavily on nonverbal cues.[5] (Countries in the Mediterranean, Central Europe, Latin America, Africa, the Middle East, and Asia fall into this category.) By contrast, explicit verbal communication is a mark of *low-context* cultures, which include most English-speaking Western cultures, including the United States and the United Kingdom.[6]

To succeed in a high-context culture, your communication must fall within traditional social and hierarchical boundaries. Work colleagues are expected to read between the lines, build and champion long-term relationships, and rely less on digital communications.

If you grew up in a high-context culture, you'll find face-to-face and phone interactions, which foster trust, to be commonplace and natural. But if you grew up, as I and other Americans did, in a low-context culture, emails and texts get to the point and can be enough to build relationships! My Italian friend Olivia always jokes about how lazy it seems when I message her with one initial O instead of *Olivia*.

Leah Johnson, a communications strategist who spent years at top posts at Citigroup and Standard & Poor, describes a common challenge when doing business in a high-context culture like Japan. "If I ask my Japanese colleagues to do something, they may not be willing at first blush to say no to me."[7] As Johnson found out, they might not say no even when they have no intention of complying. In Japanese culture, responding affirmatively when someone approaches you with a task doesn't necessarily mean you've *accepted* that task. It means only that the person understands what's been asked, or what you need. (A big impediment to Confident Collaboration for an American colleague!) To determine if the person is planning on actually *doing* what Johnson wants, she keeps her eyes open for implicit signals—silence, changing the subject, or even someone proposing an unrelated alternative to the problem. Finally understanding this, Johnson made it a habit to follow through with the appropriate decision-makers after making a request by phone (especially if she made that request in front of a group). She never, *ever* relies on email alone.

But across most English-speaking, low-context Western cultures, there's a lot less ambiguity. For example, we've all gotten emails containing attachments: *Please see attached.* Easy, right, and also efficient? But in high-context cultures like Japan and China, it can appear disrespectful to send someone a brief email that lacks contextual information or fails to acknowledge the hierarchical differences between sender and receiver.

In fact, in low-context Western cultures, people use email to communicate with practically anyone, from new employees to CEOs. It's also more common (and permissible) in low-context cultures to challenge superiors by proposing opposing views in texts or emails or during phone calls and live meetings.

Katie, the CEO of an accounting firm that does a lot of business with Chinese clients, once told me how hard it was to deal with the hierarchies of many Asian organizations. For one thing, in all communications, she was expected to cc people's managers to signal her respect for them. "If you jump one level or two to communicate beyond your reporting manager, you will be scolded for it. You should always keep your full title in the signature of your emails. The Cc field is more than 'FYI,' it is respecting the authority. Without going through or cc'ing your manager, it's common for your note to get ignored."

In high-context cultures, where Valuing Visibly is key, picking up the phone instead of sending an email is preferred even for less complex subjects. In general, digital communication is used less often in these cultures, especially during conflict mediation, idea generation, and consensus building.

Studies show that, regardless of culture, the most effective communications are direct and concise.[8] Otherwise, people reading the message or scanning it for action points and requests are likely to miss out on important details. Whether you're in a high-context or low-context culture, there are ways to modify emails to increase clarity.

In high-context cultures, begin your emails with the question you want answered, and then add a paragraph with something that personally connects you to the recipient (e.g., *How was your holiday?*). With non-native English speakers, try to avoid jargon, sports metaphors, or easily misinterpretable colloquialisms.

HOW TO COMMUNICATE
IN HIGH-CONTEXT CULTURES

- Include every detail of the business discussion.
- Ask for a response to confirm work tasks.
- Always cc a manager or ask them first before sending their report a work request.
- Include a personal non-work-related note.
- Always greet the other person before you ask for something.

HOW TO COMMUNICATE
IN LOW-CONTEXT CULTURES

- Be direct and to the point. Use bullet points and bold text to highlight the important details.
- Only say yes to a task if you plan on following through.
- Don't mix in non-work-related notes with work requests.
- Make sure it's readable on a smartphone.

If your team is US based, don't brush this chapter aside! Digital body language differences exist within America, alongside dialects and accents. East Coasters, for example, have a well-deserved reputation for directness and tend to write emails so short they can come across to non-East-Coasters as borderline unfriendly. Generally speaking, when a New Yorker says yes, they agree with you, but if they say no, feel free to argue some more! Contrast this to the West Coast, where you'll rarely hear a definite no. You're more likely to hear "I see your point" or "Let's consider some other options as well." Then there's the American South. One man I know who lives and works in North Carolina tells me he feels bulldozed when emails and phone calls don't kick off with a blizzard of charm, even if it's a simple "How are you?" By contrast, a Bostonian might sit there wondering why the person who just emailed her doesn't just get to the point.

THE SILENT HAVE SOMETHING TO SAY TOO

"The loudest duck gets shot" is an axiom many Chinese children absorb early on. If these same children grew up in a Western country, it would probably be modified to "Speak now, or forever hold your peace."

Imagine it's twenty years from now, and a person raised in China and another raised in the West are working together on the same team. Who is more likely to speak up, and who is more likely to hold their tongue?

Both online and off, silence is often a huge stumbling block to successful cross-cultural communication. Liuba Belkin, an assistant professor of management at Lehigh University, says, "In the United States, we are not very comfortable with silence. We interpret it in a very negative way. . . . [We] expend lots of useless energy searching our memories trying to recall an inadvertent slight that might have caused a friend to give us the cold shoulder. Or wonder if the latest communiqué we sent to a client was too aggressive—or not aggressive enough."[9]

As we've seen in these pages, silence in the American low-context culture—the unanswered text or email or even the length of time before the host allows you to enter a video meeting—can feel as heavy and ominous as the silence that comes after giving or receiving bad news. Contrast this with high-context cultures, where a period of silence is considered respectful, a signal that you're taking the time to reflect on what's been said and formulate the most appropriate response (silence can even be a polite way of saying no).[10]

Academia has even *studied* how silence is used across

different cultures. One 2011 bilingual analysis conducted at the University of Groningen in the Netherlands sought to measure how long it takes for people from different cultures to feel uncomfortable with conversational silence.[11] English-speaking participants lasted around four seconds in silence before they admitted they felt unsettled. Japanese-speaking participants, on the other hand, were perfectly fine with silences lasting more than double that! That's over eight seconds if you want to do the math.[12]

I once spoke to Sam, a leader managing a global remote team with members in India, the Philippines, and Cambodia. Sam had a problem. Many of the non-native English speakers on his team were silent during conference calls, and Sam had no idea what they were thinking. Accustomed to American employees sharing their thoughts haphazardly in group conversations, Sam later discovered that his Southeast Asian team members were simply unaccustomed to voicing their opinions during meetings, especially if it meant interrupting, or disagreeing with, their team leader.

In response, Sam explained that he *wanted* them to have fruitful discussions over the phone, up to and including disagreeing with him. He rearranged call agendas to make time for smaller teams in each country to speak up without them feeling as though they were disrespecting him by interjecting. He went further by asking remote team members to set goals for the number of comments and questions that they made during each call. As time went on, Sam and his global team found a workable balance—but it wasn't easy.

WHAT'S IN A NAME?

When the name "Erica" pops into your head, is it accompanied by an image? Most people who meet me virtually are surprised when they walk into a face-to-face meeting only to find an Indian woman. (In fact, most people who meet me virtually for the first time assume I'm either white, black, or biracial. My mother named me Erica to help me dodge the misinterpretations and misspellings that dogged my older sister, Darpun.)

This preconception once prompted my friend Rajesh to confess to me that he has "name envy." People are always spelling or saying his name wrong, and wouldn't it be a whole lot easier to go through life with another name? Yet when people finally hear Rajesh speak, they stumble a little. Is that an Irish accent coming out of his mouth?

If you read the name "Rajesh" and you automatically assumed "India," you might get the smallest insight into how we all perceive or prejudge certain things about one another when we've never met. Rajesh, for his part, was born in India—at least that part is right—but he grew up outside of Dublin.

Let's say you get two emails, the first from a guy named "Vinod Subramanian," and the second from a guy named "Ian Richards." Is it fair to say your mind spits out different images of their roles, titles, and even communication styles? If you say no, I don't believe you, and research backs me up. If you yourself are Indian, you are likely to develop a stronger connection to Vinod, and if you're American or British, chances are you'll feel closer to Ian.

Fact is, we *all* create unconscious biases and expectations about one another long before we meet face-to-face (if we ever do!).

. . . .

At a time when face-to-face meetings between members of the same organization are fewer and farther between than ever, recognizing our biases and predispositions represents a big step toward strengthening our workplaces.

. . . .

YOU SAY TOMATO, I SAY TOMAHTO.

Leanne, an executive, once told me about her weekly team calls for a new initiative. Her team of four people included three female native English speakers (a Brit, an American, and an Aussie) and Javier, a male from Argentina whose first language was Spanish. When Leanne instant messaged Javier to try to understand why he was being so quiet on the calls, he IM'd back, *I'm having the hardest time understanding three different English accents.* Leanne hadn't realized how hard it was for Javier, and she instituted a new rule: from now on, the group would have to slow down on calls and afterward write an email summarizing key actions and next steps. If anyone on the call needed additional clarification, Leanne requested that they email her privately.

Misunderstanding other people's accents or inflections is one thing. What about grammar and punctuation? Years ago I worked in India, and one day I got an email request from a colleague that read simply, *Please do the needful.* What did that even mean? But in Indian English, the phrase was, in fact, completely correct, and meant, "Please help me complete this task." Another phrase I saw over and over again in emails was "Let's pre-pone." The first time it happened, I emailed back, *Do you mean we need to postpone?* (I must confess, I was a little annoyed.) I later learned that "pre-pone" is a phrase used commonly in India, one that means, "We need to reschedule to an earlier date." As usual, *I* was the one who was wrong.

In my experience, the best way to preempt potential misunderstanding based on language and grammar differences is to create a psychologically safe space where those around you are comfortable letting you know when you've made a mistake. Denene Rodney, president of the cross-cultural firm Zebra Strategies, told me, "In my meetings I always say, 'I may get this wrong. If I'm saying something incorrectly, please tell me. I don't know all of the cultural differences—please make sure to correct me immediately. I look forward to learning about your culture.' Everything I do when engaging across cultures is so intentional—it's my responsibility to make others feel welcome to the table."[13]

....
**When working with other cultures, be curious,
not accusing. A question mark is better than
an exclamation point.**
....

Asking questions with a coaching mindset is a lot more effective than giving advice. As Denene told me, "Instead of saying 'Don't meet this late!' ask, 'Why are we meeting at this time?' Instead of 'I need you to do this!' say, 'Is it possible you can help me?'"[14] Communicating successfully with other cultures often involves the adaptation of so-called feminine language.

If you make a mistake—and we all do—don't simply justify yourself and move on. Apologize. "I'm sorry for my misstep. Is there a better way?" Acknowledge your mistake, and ask for clarifications and corrections. Use the opportunity to learn something new.

HI, HELLO, HEY

Greetings, signatures, and even email subject lines are the online equivalents of first and last impressions. Depending on where you're from or where you grew up, your email greetings and closings can be just as significant as your message.

Let's start with real-life greetings. Should we shake that person's hand? Air-kiss them lightly on the cheek? Give them a *double-bisou*? Should we bow, or nod? People from different cultures have different expectations about being greeted, and the same goes for emails. Bottom line: err on the side of formality during a preliminary email exchange with someone you don't know well.

Among the most common pitfalls of intercultural email etiquette is getting another person's gender wrong. (This too can show unconscious bias.) I've had my own fair share of emails from people addressing me as "Mr. Dhawan," and I only correct them when I have to. The best way of avoiding gender mishaps is to simply add a person's first name after your greeting.

In high-context cultures (such as China, India, and Turkey), use more formal language.[15] Let's say you are working with someone named Joan. A greeting like *Dear Joan* is always safe. *Joan* on its own, without a *dear* or *hello* before it, and with no punctuation in its wake, can come across as brusque or even rude. Refrain from using sarcasm or even humor.

In low-context cultures (again, like Germany, the United States, and Canada), opening an email with *Hi Joan* is perfectly appropriate. Using *hi* is, as Will Schwalbe, editor and author, put it, "perfectly friendly and innocuous."[16] *Hi* is a safe and familiar way to address someone, whether you know them or not.

But don't be surprised when you're in conversation with someone from another culture and the greeting just isn't there. Rachel, president of a PR firm, once described to me her experience working with several German clients. Many of her younger employees believed that the terseness common in Teutonic communication meant that the Germans hated them. "I had to explain that when they don't write a 'hi' or 'hey' in the greeting, they're not being rude. They're just being German!" By contrast, Rachel, herself a Gen-Xer, loved working with Germans: "They are so direct, and I like to know where I stand."

What about digital closings? Well, English email closings are likely to sound cold to Arabic speakers, who sometimes end their emails with more gracious expressions, including *Taqabalou waafir al-iHtiraam wa al-taqdeer* ("Accept an abundance of respect and appreciation"). Nigerians typically close emails with a variant of "Stay blessed." A recent comparative study by Korean and Australian academics suggests that how we close out an email has a significant effect on whether or not the recipient feels respected. In the study, 40 percent of Korean respondents found Australian emails to be impolite, compared to 28 percent the other way around.[17]

Ken Tann, a lecturer in communication management at the University of Queensland in Australia, explains, "We decide on closings based on things like familiarity and relative status. Our email closings can affect the morale and harmony of the organization, as well as our probability of getting a response. This is because it provides a way to form solidarity and the cues for our expectations on the relationship."[18]

Note too that some European countries, including Spain, France, Italy, and Portugal, typically use less formal salutations and sign-offs, whereas Americans, Germans, and Brits often begin emails with *Dear* (or simply the recipient's name), and sign off with the formal-sounding *Sincerely* or *Best*. While

Regards is an often-seen sign-off in American work emails, in the United Kingdom it can come off to some as cold. (*Kind regards* or *Best regards*, on the other hand, are both seen as warm and acceptable.)

"When I lived in the UK I thought of 'Kind regards' as fairly standard and if it got shortened to just 'Regards,' I would worry if I had offended the sender," says Leeanne Stoddart, a poet and a volunteer for several organizations in Norway.[19] She was born in the United Kingdom, but moved away as a child. "Something like 'Regards' could send me into a panic!" she added.

In the United Kingdom, ending a message with *xoxo* (an acceptable sign-off in Brazil and other Latin countries) is often deemed inappropriate. *Cheers*, which is widespread in Britain, is seldom used elsewhere, and its appearance in, say, an American email can be befuddling. Finally, the Chinese seldom bother with *any* sign-off.

As for titles, use them or lose them? In hierarchical cultures, people value seeing your formal title in your signature. When communicating with German or Japanese colleagues, ensure that you sign off with your title appearing directly beneath your name, since your status decides how quickly and carefully others respond to you. In more egalitarian cultures, there's no need to trumpet the fact that you're the "Visionary Founder" of blah-blah-blah.

BUILD BRIDGES

Understanding your own culture's communication style can help you create crucial connections across global teams. Here's how one company did it *right*.

Taimur, a 43-year-old leader in a global firm, was recently promoted to run a division whose 230 members hailed from sixteen countries, spoke nearly a dozen languages, and varied in age from 22 to 61. One of the first things he observed was the stark cultural divides between team members, which influenced their understanding of everything from the appropriate times to send emails to the best way of addressing superiors. Taimur decided to prioritize one divide in particular—the perceptions of respect and equality. Those at the company's New York office, where the boss was headquartered, believed its employees did "all the important work," whereas other, more far-flung teams, like the one based in Nairobi, felt left out. When they *did* manage to contribute something, they felt like the New York office took all the credit. As for the team in Amsterdam, they all agreed that the New York office was oblivious to what their European clients wanted and needed. The lone San Francisco team member never even got noticed.

Taimur clearly had his work cut out for him. So he got to it. First, he started embedding inclusive language such as "embrace difference" and "our shared purpose" throughout his communications. He began referring to all his teams as a single "we," creating project metrics of success that would require disparate teams to put their differences aside and work together. During his monthly calls, he highlighted how teamwork fit into every country's division-wide strategy while also giving equal airtime to each office. To ensure credit was given where it was due, he produced a monthly one-page slide highlighting the contributions of each office and their relation to the entire division. In addition, he started making phone calls or sending emails weekly across all global divisions to thank individual employees for their hard work.

SHOW EMPATHY ACROSS CULTURES

- Humanize the person on the other end of the screen by using a mixture of personal (where appropriate) and professional check-ins as tools to get to know them.
- Even if you are texting or emailing a longtime colleague or a loyal customer, be wary of using an informal tone unless they first go the more casual route.
- Avoid using abbreviations and emojis that may not translate to the other person's language or culture.

Finally, Taimur implemented norms designed to help create virtual informal connections. Teams were asked to update their email signatures with their roles and titles, include an email and conference call profile picture, and add at least one hobby to the Interests section of their company profile. Meetings involving multiple teams now began with every person announcing where they were, their specific role and focus area in the company—and even the local time. These small steps began to build greater levels of trust, familiarity, and empathy among teams, which in turn led to increased employee engagement, allowing for better work product, and creating space for cross-silo innovation like never before.

SUMMARY OF PART THREE

In part three, we explored the many ways our gender, generation, and cultural background can affect our perception of digital body language cues. Below are a set of best practices that apply to *all* these demographics.

Don't be Afraid to Discuss Differences

If you ignore gender, generational, or cultural differences, they'll simply grow in size. It's better to discuss these topics up front rather than pretend they're not there.

"When everyone is new to each other, I open our meeting by asking each team member to talk about his or her background, to make everyone aware of where their colleagues are coming from, both literally and figuratively," says Koen Bastiaens, a senior director at Cisco. "Then I try to pay attention to different ways that team members might like to engage in conversation with me and meet them where they are comfortable. It might take a one-on-one call after our meeting or an IM chat, where I ask a question less directly, to get a more accurate update on their progress." Opening up this kind of discussion mitigates the risk that some individuals might get drowned out by the louder voices on the team, and helps leaders **Value Visibly** the diversity of backgrounds on their teams.

Always Be Prepared

In my experience, the best way to fast-track a meeting, ensure results, and **Communicate Carefully** is to distribute an agenda beforehand, regardless of whether the meeting is in person, over the phone, or on video. This is especially true when working with teams of people with different levels of fluency in the language you speak. Remember that colleagues will probably vary across cultures, genders, and age groups, and that, for some, asking for clarification during a call isn't only embarrassing, it's also considered impolite in their culture. In some cases where leaders need to get feedback from lots of stakeholders, it's useful to create subteams and ask them to meet a week or two in advance to collect ideas and prepare to present the best ones to the larger group. This allows everyone to **Value Visibly.** Moreover, this configuration will prepare them to **Collaborate Confidently** with one another, regardless of any differences.

Design for Meeting Engagement

Rotate the responsibility of facilitating your meetings across time zones, age groups, and continents. For example, after two years of trying and failing to draw in greater engagement from his team, a leader based in Amsterdam asked a different team member to draft the agenda and define the group's dialogue. He did this on a weekly basis. This gave him the opportunity to hear from people he might not have heard from otherwise. A team member from Asia led the first meeting, highlighting the behind-the-scenes work completed by local colleagues. Another asked colleagues who weren't based in headquarters to be the first to share their updates. A third asked everyone present to chime in, even if they had nothing work-related to say or add.

By rotating power and driving engagement, your team is that much more likely to **Collaborate Confidently**. One good tip to ensure that everyone's on the same page and to **Value Visibly** after the meeting? Send out notes and ask for feedback via chat or even on the phone.

Even the Playing Field

Use video and interactive discussions to encourage team members to be fully present and engaged during meetings (this also prevents them from multitasking). While not everyone will get the same airtime or the same opportunity to share and defend their ideas, you can combat this inequity by insisting that everyone dial in from his or her own office. This not only prevents groups from dialing in together, it makes remote workers feel less disconnected and improves the overall effectiveness of the team. By establishing equality as an expectation from the outset, chances are good that you will build a **Trust Totally** culture.

Conclusion

When I sat down to write this book, I knew that digital body language was important. How could it not be? The majority of information we share and express today happens virtually. Yet I still insisted on framing it in my own brain as a mere comple- ment to traditional, everyday body language. I was wrong. Phys- ical body language and digital body language are inseparable. In fact, digital body language is reshaping *physical* body language, verbal communication, and even the way we *think*.

How many times have you found yourself in this scenario: A group of work colleagues sits down together for a meal. They're laughing and swapping stories, and then one of them picks up a buzzing phone, saying, "Hang on, let me just answer this text." (Similar to an alcoholic protesting, "It's *just* one drink," cell phone check-ins are always preceded by this same word, *just*: . . . *just* answering a text, *just* taking this call, *just* checking the weather forecast, *just* Shazam-ing the song playing.) A half-dozen texts later, the table has fallen silent. Since no one is talking, everyone else feels free to pull out *their* phones too. A friend of mine once dubbed cell phones "the wonder killers." Yes, it's convenient to get answers to random questions we have in real time, but the joy we lose along the way is incalculable.

Our collective addiction not only affects work culture, it also divides us. David, a retail company manager, once described his meetings with Judith, a colleague. "Whenever Judith and I met, she constantly interrupted our conversations to take another call, or respond to a text—sometimes as many as four or five times in the same meeting." David added, "It was so frustrating, especially since she found it okay. It signaled a lack of respect to me. I felt less important and less relevant."

Online and off, at our jobs or at home, our phones have altered the ways we make eye contact. We sometimes find ourselves thinking in terms of hashtags or bullet points. Our impatience levels have gone up. We expect others to get to the point fast. And nowhere is this transformation more apparent than in the workplace.

But while these downsides to digital life are real, they are balanced by more positive things. Today, digital tools can help bridge the gap between introverted and extroverted communication styles by, for example, providing extroverts with easy access to social connections while relieving introverts of lengthy extrovert-driven meetings. The digitally dependent workplace allows many of us to show what we've been made of all along and gives us the opportunity to reveal our potentials with more depth than we ever thought possible.

It's a good thing too because in the future, even the most conservative, long-established fields and businesses will be obliged to reinvent themselves digitally. I finished writing this book in quarantine with my family during the Covid-19 outbreak. As the world locked down, and many people began working from home for the first time, companies quickly adapted to Zoom, Webex, FaceTime, and—*oh, right!*—the phone. It wasn't that bad either. The number of individuals hosting Slack chats or virtual happy hours on Zoom went through the roof. We also got very good at sharing and tracking our work by email. As every organization that could entered the digital arena, it became more crucial than ever for us to master a common digital body language in order to enable clarity, speed, and efficiency.

Another thing changed as I was writing this book. My original mission was to help people alleviate a widespread source of professional confusion and pain—but along the way I learned that understanding the nuances of digital body language doesn't merely solve problems, it also opens up deeper, better ways for all of us to relate to one another and foster a sense of inclusion and belonging. This benefits *everyone* in business, from executives to managers to team members, by creating environments that allow the very best ideas to come forth and shine.

Today, we can access a much broader diversity of perspectives. Digital body language reduces friction, limits bureaucracy, and creates a clear universal language. We waste less time on low-level information. We spend more time on the things that matter. When used well, digital body language also flattens the differences across genders, generations, and cultures. Research shows that well-managed virtual teams that have never met face-to-face actually outperform co-located teams since they make use of a broader range of resources, insight, experience, and perspectives.[1] By establishing clear digital body language standards, managers now have a key tool to help them build even more successful teams in the future.

I hope this book helps you bridge the twenty-first-century differences common among teams. By helping us succeed in a modern workspace, a better understanding of digital body language goes a long way toward building greater trust, connection, and authenticity—allowing us to communicate better, build stronger relationships, and transform the way we lead, love, connect, and live.

Appendix

●

THE DIGITAL BODY LANGUAGE GUIDEBOOK

HOW DO I APPLY WHAT I'VE LEARNED?

To get you started, I've put together a practical, instructive guidebook that you can download and share with others; find it on my website at ericadhawan.com/digitalbodylanguage. You can also follow and use the hashtag #digitalbodylanguage on social media for weekly tips in action.

- **Digital Body Language Style Guide:** A digital "Elements of Style" guide that you can use alone or with your team, which includes pro tips and the dos and don'ts of digital body language.
- **Digital Styles Team Exercise:** A set of questions you can discuss with your team that demystifies digital body language style differences and helps you forge deeper understanding and trust.
- **Getting to Trust Totally:** A team facilitator guide that offers up a set of foundational basics to help you build a good digital body language culture. If you learn best by examples and reflection, this section is for you.
- **Trust Totally Quiz:** A short team exercise to help you determine trust strengths and gaps within your team to foster a Trust Totally workplace.
- **What Your Colleagues Can Tell You About Your Digital Body Language:** A short quiz that reveals your own digital communication style and what signals you may be broadcasting (even if you don't intend to!).

THE DIGITAL BODY LANGUAGE STYLE GUIDE

Want to achieve clarity in digital body language faster and more effectively? Below is a Digital Elements of Style Guide useful for any team.

- Email
- Texting and Instant Messages
- Phone or Video Meetings

EMAIL

The Audience:

❏ **Hierarchy may matter.** In some corporate cultures, it matters where people appear in the sequence of recipients. Think of the To, Cc, and Bcc lines as an old-fashioned dinner table. The boss goes at the head, and everyone else falls in line after that, depending on seniority.

❏ **Mirror the culture.** If you work in a staid, conservative culture, remember to include the appropriate formal greetings, closings, and signatures.
> Do: *Mr. Robinson,* or *Dear Sam,* . . . *Sincerely, Erica Dhawan, CEO*
> Don't: *hey Sam,* . . . *Erica*

If your work culture is more informal, use your best judgment, but make sure to mirror others appropriately.

COMMON INTERPRETATIONS OF OUR MOST USED EMAIL SIGNATURES

No signature—the equivalent of awkwardly walking out of a room, leaving everyone wondering if you accidentally hit *Send* with your elbow or if you're just ill-mannered.

Your name, no send-off—This should only be used when you are very familiar with your recipient, or you have been communicating back and forth in an email chain for more than three or four messages.

Best,—A semiformal, easy closing. This person wants you to think she's nice *and* professional. For newer relationships, opt for the more formal *Best wishes* or *Best regards.*

Regards,—Somewhat outdated, this closing is more or less neutral, but it can come across to some as distant.

Love,—Inappropriate for the workplace. Even if it's your best friend at work, *don't do it.*

Sincerely,—This formal closing is commonly used by those on a lower rung of the corporate ladder, communicating with their boss's boss. If it's not, it's probably a PR person addressing a crisis. If you don't fit into these two categories, using *Sincerely* is too formal and, in fact, may make *you* seem *insincere.*

Talk soon,—I like this one for action-oriented emails or for emails that include some kind of prep work for an upcoming meeting or phone call. It's smart, casual, practical, and friendly, but not *too* friendly.

Thanks in advance—This has actually been *shown* to be the most effective email closing around![1]

The Timing:

❏ **Email is getting faster.** A 2015 study by the USC Viterbi School of Engineering found that 50 percent of all email responses were sent within an hour. For those sent by people ages 20 to 35, that number dropped to 16 minutes. People between the ages of 35 and 50 typically answered within 24 minutes, and people 50 and older responded roughly 47 minutes later. In the time since that study was published, we are arguably only getting faster, lowering response times even further, since more of us are using our mobile phones to respond.

❏ **Value others with a "read receipt."** As email has become a faster-paced channel, it's a good idea to let the other party know that you got their email but need more than a few hours to respond. Instead of leaving the recipient waiting (or getting anxious), reply with a quick *Got it! I'll get back to you by Tuesday.*

The Structure:

❏ **Use the subject line to set the tone for your recipient before they even open your email.** Leaving a subject line empty is a wasted opportunity, and some older recipients even consider it a mark of disrespect. Consider this: how do email marketers get you to click on their message? Answer: a catchy subject line that promises you a discount, a sale, a preview, or a listicle. At work, we're all marketing ourselves—so why not get others to prioritize your requests? Do this by always starting with a specific and action-oriented subject line.

Do: Roadside Inc Project Report Final Edits / Review by EOD 4/10
Don't: Project Report

❏ **Be direct.** There's no need to restate the subject line immediately, but skip the pleasantries. For the most part, business emails don't require you to ask questions like "How's your day?" or wonder how the kids are doing. Get to the point.

❏ **Proofread for clarity, not just grammar.** *Anyone* can get confused by an email. Just because you write to co-workers whom you see every day doesn't mean they can always interpret your intent (or read your mind). Try not to be cryptic. Re-read your email and ask yourself, *If I wasn't me, or in my own head, would I understand my message?* This isn't an easy skill to develop, so ask for feedback from the receiver or from another proofreader. If a recipient responds with something other than what you wanted, ask for clarification!

Do: Let's remove the last page and reduce the total number of pages to 20
Don't: This document is too long

Pro Tips:

❏ **Be sparing with Reply All.** In general, it's only necessary if it's a one-time team notification or announcement.

❏ **Clear the deck.** If an email chain continues for more than three or four rounds, the entire subject line becomes a row of "Re:"s and "Fwd:"s. Replace those with a relevant, concise, action-oriented subject line for the email you're about to send.

❏ **Avoid stress-inducing subject lines** like *Please call me* or *See CEO in his office.* They're not what we mean by "action-oriented." Be brief, by all means, but don't forget to include the necessary context.

❏ **Embed links.** If you need to add a link, embed it into text by high-lighting the relevant words and using the insert link function offered by Office, Gmail, Yahoo!, and most other email programs. If your company works on an internal server, you can highlight the location of a document and embed it into the text in the same way.

When Should the Email Conversation Switch to a Different Medium?

❏ **Be concise.** Emails longer than five sentences are typically skimmed over! For more complex topics and assignments, opt for a phone call or meeting, or make sure you use bullet points, bold and italic text, while highlighting the action points at the end.

- **Give context.** Do I or the other person need more context beyond what email can provide? If the answer is yes, schedule a face-to-face discussion or a phone meeting.

TEXT AND INSTANT MESSAGING

Examples: text message, Skype messenger, Slack, Google Chat, etc.

As the level of formality in the workplace decreases, texting and instant messaging have become increasingly prevalent. In this section, we will combine the two shorthand forms of digital communication as their conventions are largely similar.

The Audience:

- **It's casual.** In general, your audience for these channels should be informal. The shorthand, emojis, and exaggerated punctuation used to create tone in texts and IMs are generally inappropriate in formal discussions. In a professional setting, try to write in complete sentences (though you don't have to write out every word). Abbreviations are okay.

The Timing:

- **These channels are meant to be—you guessed it—*instant.*** For the most part, a response is expected within the hour, although most come within three minutes or less. If you get a text during a meeting and can't respond immediately, it's a good idea to let recipients know why they haven't heard from you.

- **Create boundaries.** The instantaneous nature of these channels can be misleading. People often use text and IM outside of work hours, and still expect a quick response. It's perfectly reasonable to establish boundaries. If it's the first time, respond to the off-hour text with a quick message telling the sender that you won't be answering until work hours.

The Structure:

- **Less structure is better.** These are informal channels, meaning there's no reason to include a subject line, formal greeting, or signature. Too formal!

  Do: Hey, so nice to connect! Just wanted to shoot you a text so that you have my number.-Erica

Don't: Hello Stephanie, This is Erica. We met at the 2020 World Leadership Conference Dinner. I enjoyed connecting with you! Here is my phone number. Best, Erica

❏ **Get to the point.** Even more so than emails, texts and instant messages should only be used to relay information that doesn't require an in-person conversation or a phone call. Texts should be two to three sentences max.

Do: Hey Erica, are you available to meet to discuss a new project this week? Would anytime Tuesday and Thursday 1–5pm work for a 30 min call?
Don't: Dear Erica, How have you been? I'm starting a new project around collaboration in our office and thought of you. I would love to catch up.

❏ **Choose your shorthand wisely.** Only use abbreviations that are widely known *and* that you would say aloud. For example, "LMFAO" is widely known, but it's wise to avoid using it in a professional context. On the other hand, "np" (no problem) is both widely known and appropriate to use at work.

Do: np, talk soon
Don't: LMFAO yeah sure man, cya

Pro Tips:

❏ **Create a standard acronym list for common phrases.** NNTR = no need to respond, SOS = urgent, * = typo.

❏ **Don't send confidential information via text and IM!** Remember, even encrypted texts can be recorded with a screenshot.

When should an instant message conversation switch over to a different medium?

❏ Don't bait and switch. If you text or instant message someone with *hey, do you have a minute*, but then need an essay-length paragraph to explain what you need, you should probably call or email the person instead.

❏ If it's enough of an emergency to warrant a text outside of reasonable work hours (7:00 a.m.–7:00 p.m.), it probably warrants a phone call. If not, it can wait.

❏ If you need a record of the conversation, switch over to email.

VIDEO MEETINGS AND CONFERENCE CALLS

Examples: Webex, Zoom, Skype, Google Hangouts, etc.

The Audience:

❑ **Make introductions, if necessary.** Virtual meetings, especially when participants are tuning in from home, have the potential to feel more personal (and uncomfortable) than in-person meetings. Ensure that everyone is identified at the start, along with their position, and allow a few minutes for social chatter.

The Timing:

❑ **Keep things short and sweet.** Most of us are accustomed to face-to-face meetings and conference sessions that last over an hour. In that kind of three-dimensional environment, participants have a full range of stimuli, fewer outside distractions, and a longer attention span. Virtual events leave much more room for distractions and multitasking. A planned structure and a ticking clock can help. Productive virtual meetings have a predetermined time frame, which limits the number of ideas that each member will bring.

> *Do: Ask your team to bring 3 solutions to a Zoom meeting that will be capped at 60 minutes.*
> *Don't: Schedule a 3-hour Zoom meeting and forget to send out an agenda that your team can use to prepare.*

The Structure:

❑ **Raise your hand.** One great advantage of video chatting is that built-in mechanisms often exist for raising your hand (for instance, the space bar can generate hand-raising on Zoom). This can help prevent the issues common with phone calls, such as people talking over one another or cutting others off. If your software lacks a hand-raising feature, you can establish one in a side chat box. Designate a signal—like an asterisk—to allow team members to request their turn to speak. (This requires careful moderating and good leadership.) Make sure you're not overlooking anyone, and try to ask for opinions from the silent types.

❑ **Require that all participants be on camera.** As a rule, if someone else's camera is on, yours should be too. The camera restores some of the body language cues that digital communication takes away, while also allowing team members to see with their own eyes that everyone in the room is fully engaged and not scrolling Instagram.

❏ **Always appoint a moderator or MC.** Having a consistent face and voice that "stitches together" the virtual sessions for participants adds much-needed familiarity and helps to lessen feelings of isolation that may come with remote work. One good tip is to have the event moderator open up the conference or meeting before taking charge of staying on the agenda and moderating questions for speakers as they pop up in the sideline chat.

Pro Tips:

❏ **Test your technology.** If you haven't used Zoom or Skype in more than a few days, open your software before the scheduled time and test out both the video and microphone quality. This saves everybody time while allowing you to skip the "CAN YOU HEAR ME?" section of the meeting.

❏ **Silence.** Only use the mute button to prevent audio feedback and keep distractions such as breathing, writing, and fidgeting noises to a minimum. But be wary of allowing the mute button to become a license to multitask!

❏ **Slow down.** Practice the five-second rule, that is, wait five seconds before speaking after you ask the group a question. This gap allows your team to process what you just said, and it fills in the few seconds where everyone is busy wondering, *Is someone else going to speak up first?* before going forward themselves.

When should the conversation switch to a different medium?

❏ **Audit your meetings.** Just like in-person meetings, clearly define the purpose and proposed outcome of each digital meeting on your calendar and eliminate any that lack a clear purpose, or are missing a key factor for success.

DIGITAL STYLES TEAM EXERCISE

The Covid-19 pandemic provided most of us with a sustained period of learning what works and what doesn't in our own digital collaboration. Now is a good time to establish explicit norms if your team doesn't already have universal principles or rules. Each team should first consider the preferences, background, and specific functions of its members. The questions below will help team members understand their own digital body language style, which they can then share with the group. Be on the lookout for similarities and differences. These can help you establish norms and avoid potential issues.

What's My Personal Digital Body Language Style?

1. What are the best channels for others to digitally communicate with me?

2. What are my digital body language pet peeves?

3. What do I value most when others communicate with me (e.g., clarity, accountability, actions)?

4. Am I a digital adapter or a digital native? How does this "demographic" affect how I perceive daily communications?

5. Is my digital style influenced by previous work cultures or bosses? How does this show up in my communications?

List the best examples of positive digital body language I have experienced from others in each category:

Emails
Group Chat
Meetings

List the worst examples of positive digital body language I have experienced from others in each category:

Emails
Group Chat
Meetings

Have the team share their answers so everyone knows one another's preferences and peeves. Go to ericadhawan.com/digitalbodylanguage to run a full team assessment.

GETTING TO TRUST TOTALLY

Below are practices that help build the foundation necessary to achieve a culture of optimal clarity. The questions below are divided into four categories: Digital Communication, Collaboration Tools, Team Spirit, and Meeting Culture. In each section, consider the four laws we've discussed: **Value Visibly, Communicate Carefully, Collaborate Confidently,** and **Trust Totally**.

Digital Communication

The Basics

- ❏ Value Visibly: Slow down and proofread your communications as though they were presentations. Re-read what you've written, making sure your message is free of typos and confusing language. Simultaneously check for clarity, and make sure your recipient knows what you expect in response.

- ❏ Communicate Carefully: Don't rely too much on shorthand or send messages that are too general. If you want to be brief, agree on and adopt a team-wide set of acronyms to increase efficiency and clarity in digital communications.
    E.g.: WINFY: What I Need From You
    E.g.: NNTR: No Need to Respond????
    E.g.: 4H: I need this in 4 hours

- ❏ Collaborate Confidently: Assume the best intentions when you read digital communications. Remember: you are missing body language and tone cues from others, which can cause us to misinterpret directness or speed as rudeness.

Questions for Reflection

1. What are some of the recent digital communication mishaps that have happened among your team?
2. Is your team made up of more digital adapters, more digital natives, or a combo? How is this reflected in your digital communications?
3. Think of the last time that team communications made you anxious, confused, or angry. Why? Were your feelings confirmed, or was the situation a misunderstanding?
4. What is the biggest hindrance or annoyance you encounter on a daily basis when communicating digitally across the team?

Collaboration Tools

The Basics

- ❏ Communicate Carefully: Create guidelines for channel selection based on message length, expected response time, and the amount of information being transmitted. Make sure these guidelines are easily available to the entire team, especially new hires.

    E.g.: When digitally discussing sensitive client information, we only use our corporate email accounts. We do not share this type of information with each other using text messaging, IMs, or any other digital channels.

- ❏ Collaborate Confidently: Establish expectations for the appropriate timing involved in each tool, including response time, and whether certain tools are off-limits during nonworking hours.

    E.g.: Please respond to all work emails within 24 hours. If you have not received a response during that time, reach out to the recipient by phone or text.

    E.g.: Please do not send text messages regarding work tasks between the hours of 8:00 p.m. and 5:00 a.m. on any day of the week. Use your best judgment in determining whether a message is urgent enough to sidestep this guideline.

- ❏ Value Visibly: Find a handful of people who are most adept at each channel. These people can help you create guidelines and expectations, and also function as channel advocates, gently course-correcting team members who fall outside of the appropriate usage.

Questions for Reflection

1. How many collaboration tools does your team use on a daily basis?
2. Speaking for yourself, which collaboration tools personally help you thrive? Consider the potential reasons why (we've discussed many of them in this book), which can range from sheer familiarity to your preference for formality.
3. Alternately, which collaboration tools do you find yourself avoiding?
4. Is there someone in your organization who uses an X or Y collaboration tool especially well? What are they doing that you are not?
5. Does your team have an established set of norms for when each tool is to be used?
6. Which channels do you use most often as a team, and what does this say about your team culture?

Team Spirit

The Basics

- ☐ Communicate Carefully: Create spaces for informal conversations. Allow for social chatter before meetings, or create group messaging chains specifically designed for conversations that go beyond work tasks.

- ☐ Trust Totally: Create spaces for celebration. Share inspiration with your team in the form of interesting articles, podcasts, or books.

- ☐ Value Visibly: Encourage your team to show appreciation by giving each other shout-outs or starting meetings with MVP awards or "Wins of the Week." You can find your own unique way to create team spaces for social connection. How you do it is less important than whether or not you do it.

Questions for Reflection

1. Are there "cliques" among your team members? How can you bridge the gap between those groups and form a stronger bond as an entire team?
2. Is there one team member who almost always speaks the loudest? Is this person helpful or annoying? What are some of the ways you can get other voices to join in?
3. Is there one team member who is almost always the quietest? How can you encourage that person to join in?
4. How does your team deal with conflict? Are there any issues with passive-aggressive behavior? What are some tips you've learned in this book to help defuse bad behavior and/or negative team dynamics?

Meeting Culture

The Basics

- ☐ Communicate Carefully: Each meeting should be analyzed using the 5 Ps:
  - ☐ Purpose: Does the meeting have a clearly defined purpose?
  - ☐ Participants: Are all the relevant parties (and only the relevant parties) invited and able to attend?
  - ☐ Probable issues: What are the concerns that could likely arise?
  - ☐ Product: What do we want to have produced when we are done?
  - ☐ Process: What steps should we take during the meeting to achieve our purpose, given the product desired and potential issues we may face?

❏ Value Visibly: Audit your recurring meetings regularly. Every other meeting or so, ask yourself if these meetings are still necessary, if all of the appropriate players are present, and how you can improve the next one.

❏ Communicate Carefully: Make sure that someone on the team is responsible for sending out notes and action items in an email after the meeting.

❏ Collaborate Confidently: Begin each meeting with five minutes of non-task-related conversation, where participants ask about one another's days or goals or discuss with each other if and where help is needed.

❏ Trust Totally: Leaders should ensure that quieter voices are contributing to meetings by rotating leadership, asking questions, and soliciting input across various mediums.

Questions for Reflection

1. Think about your most recent meeting. Ask yourself the questions behind the 5 Ps. Are any of the answers "no"?
2. In that most recent meeting, did you feel heard and respected?
3. If not, identify the specific moments in which you felt silenced or disrespected. Can these be blamed on communication mishaps? Were you assuming the best intentions? Is it necessary to speak up on a different platform?
4. How are your meetings generally scheduled? Peer-to-peer or using assistants? Do you tend to have conversations about the need for a meeting, or do you skip ahead to simply sending a meeting request without context?
5. How can you change your scheduling process to make sure that everyone knows why the meeting exists and that the best time for everyone is chosen?

TRUST TOTALLY QUIZ

This group exercise can be a great starting point for identifying your strengths and weaknesses during a team meeting, offsite, or retreat.

Answer the following questions based on how you would respond individually within your work culture.

1. You get a calendar reminder for a meeting scheduled to begin in an hour. You are . . .
   a. Clear on why you're invited and understand what's on the agenda, and you're happy to participate.
   b. Not totally sure of why you've been invited.
   c. Befuddled why you are still on this meeting invite, and you consider skipping it or canceling at the last minute.
2. During a typical meeting, the participants . . .
   a. Are all contributing equally, following the agenda, and sharing the lead based on their areas of expertise.
   b. Are generally engaged and following the agenda if there is one. The usual loud voices take up most of the time, and their ideas are good. The manager or team lead takes charge of the entire meeting.
   c. Are getting distracted, checking emails, or answering texts. No one is following the agenda, or there isn't one. The usual loud voices are forcibly interrupting others when they speak. There's tension in the air.
3. When your manager or team lead gives out deadlines, they . . .
   a. Crowd-source your team for what's realistic by considering overall workloads and outside forces that may create delays.
   b. Set the deadline on their own, or base it on the needs of clients (or other outside forces), and present it along with the task. Usually the deadlines are realistic, but sometimes they create a time crunch. It's okay with you since you can't really control clients or suppliers.
   c. They set completely arbitrary and unrealistic deadlines. They ask for days of work to be completed overnight. You are frustrated and overworked.
4. You're having a personal issue that is affecting your concentration, so you . . .
   a. Let your team lead or manager know what's going on. You know that they will lend you support and understand any changes you may need to make to your workload.

b. Discuss it with a single trusted colleague but let no one else know. You do your best to get all of your work done and push through your day. You may have to move an important meeting to later in the week.

c. Don't tell anyone, and make no changes to your workload or schedule.

5. You've been assigned a project, but you're not sure you have the expertise to complete it, so . . .

a. You decline the project or ask that a colleague with more expertise be assigned to work on it with you.

b. You don't share your concerns with the whole team, but you know where to find the resources you need to get the task done. You may enlist a trusted colleague to help you if they can spare the time.

c. You don't share your concerns with your team even though you have no idea where to find the resources you'll need to get the task done. You end up working a 60-hour week because you're turning your wheels and doing a lot of research.

6. Your team lead shares feedback . . .

a. Often. They include kudos in each meeting, along with regular feedback on individual projects that is both useful and specific. You're not afraid when they ask to meet with you to provide feedback. You have a system of scheduled, periodic evaluations, which you typically emerge from feeling equipped with actionable tips for improvement.

b. Sometimes. They provide feedback only at the end of large projects. You get concerned when they ask to meet to give you feedback, even if you know you haven't done anything wrong. You have periodic evaluations that are seldom useful.

c. Rarely or never. You only really hear feedback if something goes wrong. Your anxiety levels would soar if they asked you to meet for a feedback session.

7. Your organization's evaluations are . . .

a. 360 degrees. Everyone on the team has an opportunity to hear from all levels. Leaders and executives get feedback from their subordinates. Peers are able to give each other feedback as well.

b. A back-and-forth. Team leads get feedback from their subordinates. Subordinates get feedback from their team leads. Peers are not encouraged to share feedback with each other.

      c. One-sided. Team leads give feedback to their subordinates, but do not receive any. Peers are not encouraged to share feedback with one another.

8. When it comes to channel selection, your team . . .

      a. Has a clear set of norms regarding message length, expected response time, and the amount of information being transmitted. You never wonder which channel to use or get annoyed if another person uses the wrong one.

      b. Doesn't have a clear set of norms—but there isn't much confusion about it either. Usually you're not concerned about which channel to use, but now and then you get annoyed by out-of-the-blue phone calls, confusing or cryptic emails, or inappropriate texts and IMs.

      c. Is a mess. You don't have any norms and there is often confusion. Messages are often chronically lost, or never get answered.

9. It's the night before a major presentation and your partner hasn't added their slides to the PowerPoint. You . . .

      a. Aren't too worried. You're sure they'll get it done. Just to be safe, you send a text that reads, *Hey, just checking in! My slides are all ready to be integrated with yours. Excited for tomorrow!*

      b. Are a bit worried that maybe they have the date wrong, or even forgot about it. You send them a text that says, *Hey, any idea when you'll be done with the project? Just wanted to be able to run through it before tomorrow's presentation . . .*

      c. Panic. You know they probably forgot, or else they expect you to do it yourself. You call them, and when they don't answer, you proceed to just do it yourself.

10. You feel like a valuable part of your team . . .

      a. All the time. You're asked to contribute regularly, and you feel comfortable sharing opinions and making suggestions. Your teammates and leaders often give you kudos for your contributions.

      b. When your area of expertise is front and center. You try to stay quiet when you're not 100 percent sure of your opinions. When you do speak up, your team acknowledges your good ideas.

      c. Rarely or never. You hold your tongue as much as possible, and complete your tasks narrowly. You seldom get recognition from the other members of your team for your work.

If your scores are mostly A's: Congratulations, you are close to being aligned across your team! Focus on the gaps identified in the questions.

If your scores are mostly B's: Your team is doing fine, but you can do better. Focus on the Digital Body Language Style Guide.

If your scores are mostly C's: Your team needs a lot of help. Jump to Getting to Trust Totally to get started.

Now, review your answers across the team. What you'll quickly notice is that team member experiences of clarity vary widely. Often, the leader of a team will have higher scores than their peers, or some silos will have higher scores than others. Identify the asymmetries in individual scores and discuss why they occur.

WHAT YOUR COLLEAGUES CAN TELL ABOUT YOUR DIGITAL BODY LANGUAGE

This is a fun, quick, and incredibly useful way to get feedback, and it will help you assess the digital body language signals that you may or may not realize you're projecting. Ask a colleague to identify which person best represents you: Alice, Betty, Charlie, or David.

Alice

The emails that Alice sends are short and to the point, but she always takes the time to craft a specific and useful subject line and proofread her emails for clarity. She may include a single pleasantry—*Have a great day!* or *Let me know if I can help further.* She uses emojis as additions to texts and IMs. They don't take the place of words, but she'll usually add a smiley face or a thumbs-up to the end of a message to add some emotion. Generally Alice answers emails within two or three hours, responds to texts within a few minutes, and answers calendar requests almost immediately. If she knows her response will be delayed, she always lets the other person know. In her team communication, she follows the norms in all communication guidelines, honoring issues around length, complexity, and familiarity.

Betty

Betty's emails are always polite. Even if she's apologetic about something, you'll never know it. Her most used emoji is 😊. As for response times, it depends on the person who sent her the email. If it's her boss, she'll respond at once. If it's from someone she dislikes, she'll push against the boundaries of acceptable (but never too far).

Charlie

The emails Charlie sends always contain fewer than five sentences. It's like texting. You and Charlie tend to go back and forth using short emails. Charlie loves emojis, as a single image is much easier to type than a full sentence, and he easily substitutes emojis for words or even sentences. He is more than willing to overlook a few typos here and there for the sake of speed.

David

David sends lengthy emails that overflow with details, negating any need for a follow-up email or call. Sometimes his emails contain several paragraphs, bulleted lists, links, and attachments—whatever's necessary. He would never use emojis at work, as he sees them as unprofessional (plus, to be honest, he has no idea what some of them mean). He always triple-checks his messages before pressing *Send*.

If you are mostly Alice: Great job. The foundations of your communications are solid. How are you using it to propel your career forward?

If you are mostly Betty: Depending on your work environment, your communications may come across as passive-aggressive or confusing. Work on these tips:

- **Value Visibly:** Remember to express appreciation with a simple "thanks" or by letting someone know when they've done a good job.
- **Communicate Carefully:** Avoid sending messages when you're angry or frustrated.
- **Collaborate Confidently:** Be direct about what you need and what you are feeling.

If you are mostly Charlie: You may be sacrificing clarity for speed and efficiency. Work on these tips:

- **Communicate Carefully:** Slow down. Ask yourself: Is it clear what the recipient needs to do, why, and by when?
- **Collaborate Confidently:** Avoid short, anxiety-inducing messages like *We need to talk* or *That might work*.

If you are mostly David: You may be sending messages that are too complex and, therefore, unclear: Try these tips:

- **Communicate Carefully:** Review when it's better to switch mediums. Bear in mind, too, that complexity is also a factor for channel selection.
- **Collaborate Confidently:** Get comfortable on phone and video calls! Sometimes we simply have too much to say, and it's easier to do so using a channel where the meaning of our words is enhanced by our tone and our ability to ask questions as they come up.

Acknowledgments

There are many incredible people who helped me bring this book to life. My literary guide, and agent, Jim Levine, who's been on this journey with me from the very beginning. I am deeply grateful for my powerhouse publishers at St. Martin's Press: Tim Bartlett, Alice Pfeifer, Laura Clark, Rebecca Lang, Danielle Prielepp, Alan Bradshaw, and the entire Macmillan family.

Thank you Abby Salinas, you are far more than my cowriter, you are my dear friend. Thank you to Peter Smith and Lily Smith for helping me ideate and bring this book to life. Thank you to Omeed Rameshni for serving as my trusted advisor for many years.

A big shout-out to my husband, Rahul—you are my best friend, soulmate, and forever confidant. Thank you to my parents, Ram and Neelam Dhawan, for teaching me the importance of grit, empathy, and generosity. Thank you to my siblings, Neil and Darpun, for showing excellence in their own fields and inspiring me to do the same.

Special thank-you to my many close advisers and supporters: Rob Berk, Ayse Birsel, Shelley Brindle, Irene Britt, Bill and Julie Carrier, Meg Cassidy, Alisa Cohn, Lin Coughlin, Dorie Clark, Laine Cohen, Adette Contreras, Mark Fortier, Marshall Goldsmith, Patricia Gorton, Amanda Hughey, Leah Johnson, Mo Kasti, Carrie Kerpen, Randi Kochman, Stephanie Land, Alex Lapshin, Ivy Lin, Amanda Hughey, Emily Mills, Will Morel, Scott Osman, Michael Palgon, Jordan Pople, Denene Rodney, Rajeev Ronanki, Brad Schiller, Lisa Santandrea, Lisa Shalett, Kim Sharan, Dan Schawbel, Maesha Shonar, Darcy Verhun, and Leslie Zaikis.

This book was finished during the Covid-19 pandemic. It's worth taking a second to think about people who did more than expected, more than they had to do, more than we can imagine in unprecedented times. I'm filled with gratitude for the healthcare workers who showed up to do the jobs that they never hoped to have to do, risking so much to help our communities. Thanks to everyone I met and shared with on Zoom.

I'm inspired by the team at Cotential, working remotely to deliver important client work and connecting our extraordinary community of tens of thousands of people.

And I'm grateful to you, loyal reader, for taking the long view, for leading, for spreading the skills of digital body language and showing up and connecting with others in ways that you're proud of. You are my hero.

Notes

Introduction

1. Marguerite Ward, "CEO of a $16 Billion Business Says the Way You Write Emails Can Break Your Career," CNBC Careers, November 30, 2016, https://www.cnbc.com/2016/11/30/ceo-of-a-16-billion-business-says-the-way-you-write-emails-can-break-your-career.html
2. Ibid.

Chapter 1: What *Is* Digital Body Language?

1. The Radicati Group, Inc., "Email Statistics Report, 2020–2024," February 2020, https://www.radicati.com/wp/wp-content/uploads/2019/12/Email-Statistics-Report-2020-2024-Executive-Summary.pdf; The Radicati Group, Inc., "Email Statistics Report, 2015–2019," March 2015, https://www.radicati.com/wp/wp-content/uploads/2015/02/Email-Statistics-Report-2015–2019-Executive-Summary.pdf
2. Justin Kruger, Nicholas Epley, Jason Parker, and Zhi-Wen Ng, "Egocentrism over E-mail: Can We Communicate as Well as We Think?" *Journal of Personality and Social Psychology* 89, no. 6 (December 2005): 925–36, https://doi.org/10.1037/0022–3514.89.6.925
3. Niraj Chokshi, "Out of the Office: More People Are Working Remotely, Survey Finds," *New York Times*, February 15, 2017, https://www.nytimes.com/2017/02/15/us/remote-workers-work-from-home.html
4. Annalise Knudson, "Teens Prefer Texting to Talking, New Survey Shows," silive, September 11, 2018, https://www.silive.com/news/2018/09/teens_prefer_texting_to_talkin.html
5. Lee Rainie and Kathryn Zickuhr, "Americans' Views on Mobile Etiquette,"

Pew Research Center, December 31, 2019, https://www.pewresearch.org/internet/2015/08/26/americans-views-on-mobile-etiquette/

6. Carolyn Sun, "How Do Your Social Media Habits Compare to the Average Person's?" *Entrepreneur*, December 14, 2017, https://www.entrepreneur.com/slideshow/306136

7. Allan Pease and Barbara Pease, *The Definitive Book of Body Language* (New York: Bantam Books, 2006), 10.

8. Katrin Schoenenberg, Alexander Raake, and Judith Koeppe, "Why Are You So Slow? Misattribution of Transmission Delay to Attributes of the Conversation Partner at the Far-End," *International Journal of Human-Computer Studies* 72, no. 5 (May 2014): 477–87, https://doi.org/10.1016/j.ijhcs.2014.02.004

Chapter 2: Why Are You *So* Stressed?

1. *The Devil Wears Prada*, directed by David Frankel (Beverly Hills, CA: 20th Century Fox Home Entertainment, 2006).

2. John Suler, "The Online Disinhibition Effect," *CyberPsychology & Behavior* 7, no. 3 (2004): 321–26, https://doi.org/10.1089/1094931041291295

3. Keith Ferrazzi, "How to Avoid Virtual Miscommunication," *Harvard Business Review*, March 31, 2020, https://hbr.org/2013/04/how-to-avoid-virtual-miscommun

4. Alina Dizik, "How to Avoid Writing Irritating Emails," BBC Worklife, September 5, 2017, https://www.bbc.com/worklife/article/20170904-how-to-avoid-writing-irritating-emails

5. Ibid.

6. Gareth Cook, "The Secret Language Code," *Scientific American*, August 16, 2011, https://www.scientificamerican.com/article/the-secret-language-code/

7. Eugene Wei, "Pronoun Usage: A Psychological Tell," Remains of the Day, August 26, 2011, https://www.eugenewei.com/blog/2011/8/26/pronoun-usage-a-psychological-tell.html

Chapter 3: What Are You *Really* Saying?

1. Katrin Schoenenberg, "Awkward Pauses in Online Calls Make Us See People Differently," *The Conversation*, April 2, 2020, https://theconversation.com/awkward-pauses-in-online-calls-make-us-see-people-differently-26073

2. Alisha Haridasani Gupta, "It's Not Just You: In Online Meetings, Many Women Can't Get a Word In," *New York Times*, April 14, 2020, https://www.nytimes.com/2020/04/14/us/zoom-meetings-gender.html

3. Jessica Stillman, "A Simple Way to Make Conference Calls Less Awkward," *Inc.*, November 18, 2014, https://www.inc.com/jessica-stillman/the-5-second-secret-to-less-awkward-online-meetings.html

4. *Seinfeld*, season 5, episode 4, "The Sniffing Accountant."

5. Aimee Lee Ball, "Talking (Exclamation) Points," *New York Times*, July 3, 2011, http://www.nytimes.com/2011/07/03/fashion/exclamation-points-and-e-mails-cultural-studies.html

6. Emily Torres, "The Danger of Overusing Exclamation Marks," BBC Worklife, June 7, 2019, https://www.bbc.com/worklife/article/20190606-the-danger-of-overusing-exclamation-marks

7. Tanya Dua, "Emojis by the Numbers: A Digiday Data Dump," *Digiday*, May 8, 2015, http://digiday.com/marketing/digiday-guide-things-emoji/; Uptin Saiidi,

"Brand Marketers Find a New Way into Your Phone," CNBC, August 19, 2014, https://www.cnbc.com/2014/08/18/emojis-brand-marketers-find-a-new-way-into-your-phone.html

8. Samantha Lee, "What Communicating Only in Emoji Taught Me About Language in the Digital Age," *Quartz*, August 31, 2016, https://qz.com/765945/emojis-forever-or-whatever-im-a-poet/

9. "#CHEVYGOESEMOJI," Chevrolet Pressroom, June 22, 2015, https://media.chevrolet.com/media/us/en/chevrolet/news.detail.html/content/Pages/news/us/en/2015/jun/0622-cruze-emoji.html

10. Eric Goldman and Gabriella Ziccarelli, "How a Chipmunk Emoji Cost an Israeli Texter $2,200," *Technology & Marketing Law Blog*, May 25, 2017, https://blog.ericgoldman.org/archives/2017/05/how-a-chipmunk-emoji-cost-an-israeli-texter-2200.htm

11. Ella Glikson, Arik Cheshin, and Gerben A. Van Kleef, "The Dark Side of a Smiley: Effects of Smiling Emoticons on Virtual First Impressions," *Social Psychological and Personality Science* 9, no. 5 (July 31, 2017): 614–25, https://doi.org/10.1177/1948550617720269

12. Alice Robb, "How Using Emoji Makes Us Less Emotional," *The New Republic*, July 7, 2014, https://newrepublic.com/article/118562/emoticons-effect-way-we-communicate-linguists-study-effects

13. Danielle N. Gunraj, April M. Drumm-Hewitt, Erica M. Dashow, Sri Siddhi N. Upadhyay, and Celia M. Klin, "Texting Insincerely: The Role of the Period in Text Messaging," *Computers in Human Behavior* 55, pt. B (February 2016): 1067–75, https://doi.org/10.1016/j.chb.2015.11.003

14. Paige Lee Jones, Twitter, December 21, 2017, 1:39 p.m., https://twitter.com/paigeleejones/status/943928863163371520

15. Ed Yong, "The Incredible Thing We Do During Conversations," *The Atlantic*, January 4, 2016, https://www.theatlantic.com/science/archive/2016/01/the-incredible-thing-we-do-during-conversations/422439/

16. "The Wireless Industry: Industry Data," CTIA, https://www.ctia.org/the-wireless-industry/infographics-library (accessed April 2, 2020).

17. Alina Tugend, "The Anxiety of the Unanswered E-Mail," *New York Times*, April 20, 2013, https://www.nytimes.com/2013/04/20/your-money/the-anxiety-of-the-unanswered-e-mail.html

Chapter 4: Value Visibly

1. Christine Porath, "Half of Employees Don't Feel Respected by Their Bosses," *Harvard Business Review*, December 6, 2017, https://hbr.org/2014/11/half-of-employees-dont-feel-respected-by-their-bosses

2. Nicole Spector, "Why Are Big Companies Calling Their Remote Workers Back to the Office?" NBCNews.com, July 27, 2017, https://www.nbcnews.com/business/business-news/why-are-big-companies-calling-their-remote-workers-back-office-n787101

3. Naomi S. Baron, *Words Onscreen: The Fate of Reading in a Digital World* (New York: Oxford University Press, 2016), 168.

4. Ibid.

5. Scott Gerber (CEO, Young Entrepreneur Council), in discussion with the author, May 2019.

6. Aria Finger (CEO, DoSomething.org), in discussion with the author, May 2019.

7. "NPT's Best Nonprofits to Work For 2013," *The NonProfit Times*, April 1, 2013;

"Best Places to Work—DOSOMETHING.ORG," Crain's New York Business, January 1, 2012, https://www.crainsnewyork.com/awards/dosomethingorg-3

8. Sara Algoe, "Putting the 'You' in Thank You," *Journal of Social Psychology*, June 7, 2016, https://journals.sagepub.com/doi/10.1177/1948550616651681

9. Jena McGregor, "The Odd Things People do While Half-Listening on Conference Calls," *Washington Post*, August 21, 2014, https://www.washingtonpost.com/news/on-leadership/wp/2014/08/21/the-odd-things-people-do-while-half-listening-on-conference-calls/

Chapter 5: Communicate Carefully

1. Alon Schwartz, "Does Your Team Know What Success Looks Like?" alonshwartz.com, March 20, 2019, http://alonshwartz.com/2019/03/does-your-team-know-what-success-looks-like/

2. Ibid.

3. Ibid.

4. "Poor Communication Leads to Project Failure One Third of the Time," Coreworx, April 20, 2017, https://info.coreworx.com/blog/pmi-study-reveals-poor-communication-leads-to-project-failure-one-third-of-the-time

5. "The High Cost of Low Performance: The Essential Role of Communications," Project Management Institute, May 2013, https://www.pmi.org/-/media/pmi/documents/public/pdf/learning/thought-leadership/pulse/the-essential-role-of-communications.pdf

6. Daniel Victor and Matt Stevens, "United Airlines Passenger is Dragged from an Overbooked Flight," *New York Times*, April 10, 2017, https://www.nytimes.com/2017/04/10/business/united-flight-passenger-dragged.html

7. Lucinda Shen, "United Airlines Stock Drops $1.4 Billion After Passenger-Removal Controversy," *Fortune*, April 11, 2017, https://fortune.com/2017/04/11/united-airlines-stock-drop/

8. Liam Stack and Matt Stevens, "Southwest Airlines Engine Explodes in Flight, Killing a Passenger," *New York Times*, April 17, 2018, https://www.nytimes.com/2018/04/17/us/southwest-airlines-explosion.html

9. "Southwest Flight Suffers Jet Engine Failure," CNN, April 17, 2018, https://www.cnn.com/us/live-news/southwest-flight-emergency/h_e24cbf88f32766bb168d5bafd6539538

10. WFAA, "Southwest CEO Mourns Loss of Passenger on Dallas-Bound Flight," *YouTube* (video), 18:20, April 17, 2018, https://www.youtube.com/watch?v=hu3yfAA8aI8

11. Wade Foster et al., *The Ultimate Guide to Remote Work (ebook)*, ed. Danny Schreiber and Matthew Guay (Sunnyvale, CA: Zapier, 2015), https://cdn.zapier.com/storage/learn_ebooks/e4fbeb81f76c0c13b589cd390cb6420b.pdf

12. Ibid.

13. Ibid.

Chapter 6: Collaborate Confidently

1. Tom Monahan, "The Hard Evidence: Business is Slowing Down," *Fortune*, January 28, 2016, https://fortune.com/2016/01/28/business-decision-making-project-management/

2. Ibid.

3. Ibid.
4. Françoise Henderson, "Translating Your Product for the Global Market? Beware the Silo Effect," *Global Trade Mag*, January, 17, 2020, https:// www.globaltrademag.com/translating-your-product-for-the-global-market -beware-the-silo-effect/?gtd=3850&scn=
5. Ibid.
6. Ibid.
7. "The 2013 Regulatory Landscape from FinanceConnect:13," May 9, 2013, https://www.youtube.com/watch?v=cbsMDRDBB_o
8. "The Bloody History of 'Deadline,'" Merriam-Webster, https://www.merriam -webster.com/words-at-play/your-deadline-wont-kill-you (accessed April 4, 2020).

Chapter 7: Trust Totally

1. Amy Feldman, "Away Luggage Hits $1.4B Valuation After $100M Fundraise," *Forbes*, May 15, 2019, https://www.forbes.com/sites/amyfeldman/2019/05 /14/at-a-valuation-as-high-as-145b-valuation/#2a5fc8dc33d7
2. Ingrid Angulo, "Facebook and YouTube Should Have Learned from Micro- soft's Racist Chatbot," CNBC, March 17, 2018, https://www.cnbc.com /2018/03/17/facebook-and-youtube-should-learn-from-microsoft-tay-racist -chatbot.html
3. Peter Lee, "Learning from Tay's Introduction," *Official Microsoft Blog*, March 25, 2016, https://blogs.microsoft.com/blog/2016/03/25/learning-tays-introduction /#sm.0000x5ncvafjkel7qin1ue35ompd9
4. Justin Bariso, "Microsoft's CEO Sent an Extraordinary Email to Employees After They Committed an Epic Fail," *Inc.*, February 23, 2017, https://www.inc.com /justin-bariso/microsofts-ceo-sent-an-extraordinary-email-to-employees -after-they-committed-an-.html
5. Marco della Cava, "Microsoft's Satya Nadella is Counting on Culture Shock to Drive Growth," *USA Today*, February 20, 2017, https://www.usatoday.com /story/tech/news/2017/02/20/microsofts-satya-nadella-counting-culture -shock-drive-growth/98011388/
6. Shana Lebowitz, "Google Considers This to Be the Most Critical Trait of Suc- cessful Teams," *Business Insider Australia*, November 21, 2015, https://www .businessinsider.com.au/amy-edmondson-on-psychological-safety-2015–11
7. Amy Edmondson, "Psychological Safety and Learning Behavior in Work Teams," *Administrative Science Quarterly* 44, no. 2 (June 1999): 350–83, https://doi.org/10.2307/2666999

Chapter 8: Gender

1. John Paul Titlow, "These Women Entrepreneurs Created a Fake Male Co- founder to Dodge Startup Sexism," *Fast Company*, September 19, 2017, https://www.fastcompany.com/40456604/these-women-entrepreneurs -created-a-fake-male-cofounder-to-dodge-startup-sexism
2. Ibid.
3. John Gray, *Men Are from Mars, Women Are from Venus: The Classic Guide to Understanding the Opposite Sex* (New York: Harper, 2012).
4. Leah Fessler, "Your Company's Slack is Probably Sexist," *Quartz*, November 20, 2017, https://qz.com/work/1128150/your-companys-slack-is-probably-sexist/

5. Ibid.
6. Daniel N. Maltz and Ruth A. Borker, "A Cultural Approach to Male-Female Miscommunication," in *Language and Social Identity*, ed. John J. Gumperz, *Studies in Interactional Sociolinguistics* (Cambridge: Cambridge University Press, 1983), 196–216, doi:10.1017/CBO9780511620836.013
7. Ibid.
8. Ibid.
9. "Overview," Project Implicit, 2011, https://implicit.harvard.edu/implicit /education.html
10. Leah Fessler, "Your Company's Slack is Probably Sexist," *Quartz*, November 20, 2017, https://qz.com/work/1128150/your-companys-slack-is-probably -sexist/
11. Margarita Mayo, "To Seem Confident, Women Have to be Seen as Warm," *Harvard Business Review*, November 26, 2019, https://hbr.org/2016/07/to -seem-confident-women-have-to-be-seen-as-warm
12. James Fell, "I just said this to my feminist wife and daughter and they both laughed and agreed, so I don't think it qualifies as sexist," Facebook, February 6, 2017, https://www.facebook.com/bodyforwife/posts/1330446537016138
13. James Fell, "Why Men Don't Use Exclamation Points (and Women Do)," James Fell (blog), July 25, 2019, https://bodyforwife.com/why-men-dont -use-exclamation-points-and-women-do/
14. Ibid.
15. Naomi S. Baron, *Always On: Language in an Online and Mobile World* (Oxford: Oxford University Press, 2010), 52.
16. Naomi Baron, "Dr. Naomi Baron: Maximizing and Using Digital Communication Skills in Leadership, Episode #21," *Masters of Leadership with Erica Dhawan* (podcast), May 1, 2018, https://ericadhawan.com/dr-naomi-baron -maximizing-and-using-digital-communication-skills-in-leadership-episode -21/
17. Leah Fessler, "Your Company's Slack is Probably Sexist," *Quartz*, November 20, 2017, https://qz.com/work/1128150/your-companys-slack-is-probably -sexist/
18. Ibid.
19. Ibid.
20. Anil Dash, "The Year I Didn't Retweet Men," Medium, February 13, 2014, https://medium.com/the-only-woman-in-the-room/the-year-i-didnt-retweet -men-79403a7eade1
21. Ibid.
22. Ibid.
23. Ibid.
24. Mark Peters, "The Hidden Sexism in Workplace Language," BBC Worklife, March 30, 2017, https://www.bbc.com/worklife/article/20170329-the -hidden-sexism-in-workplace-language
25. Maxwell Huppert, "5 Must-Do's for Writing Inclusive Job Descriptions," *LinkedIn Talent Blog*, LinkedIn, April 9, 2018, https://business.linkedin .com/talent-solutions/blog/job-descriptions/2018/5-must-dos-for-writing -inclusive-job-descriptions
26. Samantha Cole, "How Changing One Word in Job Descriptions Can Lead to More Diverse Candidates," *Fast Company*, March 24, 2015, https://www

.fastcompany.com/3044094/how-changing-one-word-in-job-descriptions
-can-lead-to-more-diverse-candid

27. Yoree Koh, "How Language in Job Listings Could Widen Silicon Valley's Gender Divide," *Wall Street Journal*, December 13, 2017, https://www.wsj.com/articles/how-language-in-job-listings-could-widen-silicon-valleys-gender-divide-1513189821

28. Kieran Snyder, "Language in Your Job Post Predicts the Gender of Your Hire," Textio, June 21, 2016, https://textio.com/blog/language-in-your-job-post-predicts-the-gender-of-your hire/13034792944

29. Tim Halloran, "Better Hiring Starts with Smarter Writing," Textio, June 16, 2017, https://textio.com/blog/better-hiring-starts-with-smarter-writing/13035166297

30. Noam Scheiber and John Eligon, "Elite Law Firm's All-White Partner Class Stirs Debate on Diversity," *New York Times*, January 27, 2019, https://www.nytimes.com/2019/01/27/us/paul-weiss-partner-diversity-law-firm.html

31. Cassens Weiss, "170 Top In-House Lawyers Warn They Will Direct Their Dollars to Law Firms Promoting Diversity," *ABA Journal*, American Bar Association, January 28, 2019, https://www.abajournal.com/news/article/170-top-in-house-lawyers-warn-they-will-direct-their-dollars-to-law-firms-promoting-diversity

32. Leah Fessler, "Your Company's Slack is Probably Sexist," *Quartz*, November 20, 2017, https://qz.com/work/1128150/your-companys-slack-is-probably-sexist/

33. Ibid.

34. Ibid.

35. Ibid.

36. Susan C. Herring, "Communication Styles Make a Difference," The Opinion Pages, *New York Times*, February 4, 2011, https://www.nytimes.com/roomfordebate/2011/02/02/where-are-the-women-in-wikipedia/communication-styles-make-a-difference

37. Ibid.

38. Sirin Kale, "Working from Home? Video Conference Call Tips for the Self-Isolating," *The Guardian*, March 14, 2020, https://www.theguardian.com/money/2020/mar/14/video-conference-call-tips-self-isolating-coronavirus-working-from-home

Chapter 9: Generation

1. Jackie L. Hartman and Jim McCambridge, "Optimizing Millennials' Communication Styles," *Business Communication Quarterly* 74, no. 1 (February 23, 2011): 22–44, https://doi.org/10.1177/1080569910395564

2. Ibid.

3. Meghan McCarty Carino, "Ellipses and Emoji: How Age Affects Communication at Work," *Marketplace*, Minnesota Public Radio, October 23, 2019, https://www.marketplace.org/2019/10/21/ellipses-and-emoji-how-age-affects-communication-at-work/

4. Jenna Goudreau, "How to Communicate in the New Multigenerational Office," *Forbes*, February 14, 2013, https://www.forbes.com/sites/jennagoudreau/2013/02/14/how-to-communicate-in-the-new-multigenerational-office/#2e62918e4a6b

5. Meghan McCarty Carino, "Ellipses and Emoji: How Age Affects Communication at Work," *Marketplace*, Minnesota Public Radio, October 23, 2019, https://www.marketplace.org/2019/10/21/ellipses-and-emoji-how-age-affects-communication-at-work/

6. Christopher Mims, "Yes, You Actually Should be Using Emojis at Work," *Wall Street Journal*, July 20, 2019, https://www.wsj.com/articles/yes-you-actually-should-be-using-emojis-at-work-11563595262

7. Jay Reeves, "Five Tips for Using Emojis Without Getting Sued," *Byte of Prevention Blog*, Lawyers Mutual, April 9, 2020, https://www.lawyersmutualnc.com/blog/five-tips-for-using-emojis-without-getting-sued

8. Rachel Been, Nicole Bleuel, Agustin Fonts, and Mark Davis, "Expanding Emoji Professions: Reducing Gender Inequality," Google LLC, May 11, 2016, https://unicode.org/L2/L2016/16160-emoji-professions.pdf

9. Jazmine Hughes, "Need to Keep Gen Z Workers Happy? Hire a 'Generational Consultant,'" *New York Times Magazine*, February 19, 2020, https://www.nytimes.com/interactive/2020/02/19/magazine/millennials-gen-z-consulting.html

Chapter 10: Culture

1. Olga Khazan, "The Countries Where Smiling Makes You Look Dumb," *The Atlantic*, May 27, 2016, https://www.theatlantic.com/science/archive/2016/05/culture-and-smiling/483827/

2. Echo Huang, "Chinese People Mean Something Very Different When They Send You a Smiley Emoji," *Quartz*, March 29, 2017, https://qz.com/944693/chinese-people-mean-something-very-different-when-they-send-you-a-smiley-emoji/

3. Alex Rawlings, "Why Emoji Mean Different Things in Different Cultures," BBC Future, December 11, 2018, https://www.bbc.com/future/article/20181211-why-emoji-mean-different-things-in-different-cultures

4. Ryan Holmes, "Are You Using the Wrong Emojis at Work?" *Forbes*, July 16, 2019, https://www.forbes.com/sites/ryanholmes/2019/07/16/are-you-using-the-wrong-emojis-at-work/#5dcd42252c42

5. Arhlene A. Flowers, *Global Writing for Public Relations: Connecting in English with Stakeholders and Publics Worldwide* (New York: Routledge, Taylor & Francis Group, 2016), 255.

6. Ibid.

7. Erica Dhawan, "How to Create a Culture of Collaboration," *Forbes*, January 27, 2017, https://www.forbes.com/sites/ericadhawan/2017/01/27/how-to-create-a-culture-of-collaboration/#39f710e133fe

8. Eric Barton, "Master the Art of Global Email Etiquette," BBC Worklife, November 7, 2013, https://www.bbc.com/worklife/article/20131106-lost-in-translation

9. Alina Tugend, "The Anxiety of the Unanswered E-mail," Business Shortcuts, *New York Times*, April 19, 2013, https://www.nytimes.com/2013/04/20/your-money/the-anxiety-of-the-unanswered-e-mail.html

10. John Hooker, "Cultural Differences in Business Communication," Tepper School of Business, Carnegie Mellon University, December 2008, https://public.tepper.cmu.edu/jnh/businessCommunication.pdf

11. Lennox Morrison, "The Subtle Power of Uncomfortable Silences," BBC

Worklife, July 18, 2017, https://www.bbc.com/worklife/article/20170718 -the-subtle-power-of-uncomfortable-silences

12. Ibid.
13. Denene Rodney (president, Zebra Strategies), in discussion with the author, May 2019.
14. Ibid.
15. Arhlene A. Flowers, *Global Writing for Public Relations: Connecting in English with Stakeholders and Publics Worldwide* (New York: Routledge, Taylor & Francis Group, 2016), 258.
16. Business Insider, "These Are the Best and Worst Ways to Start an Email," *Fortune*, August 10, 2017, https://fortune.com/2017/08/10/email-etiquette -best-worst-start/
17. Margaret Murphy and Mike Levy, "Politeness in Intercultural Email Communication: Australian and Korean Perspectives," *Journal of Intercultural Communication* 12(2006), https://www.immi.se/intercultural/nr12 /murphy.htm
18. Christine Ro, "The Beautiful Ways Different Cultures Sign Emails," BBC Worklife, May 10, 2019, https://www.bbc.com/worklife/article/20190508 -why-the-way-you-close-your-emails-is-causing-confusion
19. Ibid.

Conclusion

1. Keith Ferrazzi, "Virtual Teams Can Outperform Traditional Teams," *Harvard Business Review*, March 20, 2012, https://hbr.org/2012/03/how-virtual -teams-can-outperfo

Appendix

1. Adam M. Grant and Francesca Gino, "A Little Thanks Goes a Long Way: Explaining Why Gratitude Expressions Motivate Prosocial Behavior," *Journal of Personality and Social Psychology* 98, no. 6 (2010): 946–55, https://doi.org /10.1037/a0017935

Index

emails and, 29, 34–39, 137
fear and, 130
feelings behind common phrases, 37
microaggression and, 35
punctuation and, 66
Trust and Power Matrix and, 38
Paul, Weiss (law firm), 183
Pennebaker, James, 44
period (punctuation), 11, 47, 57, 64–65,
175, 188. *See also* punctuation
phone calls
choosing a communication medium,
21, 29, 32, 34, 37–38, 45, 48,
49–52, 54
complexity and, 120
generational differences and, 190,
193–195
respect for time and, 70–73
stressors and, 24
Value Visibly and, 90, 91, 94–95, 97,
100–101
See also voicemail
phone-phobia, 51
Plank, Elizabeth, 167
posture, body, xiv, 8, 46, 189
power
formality and, 42, 43, 44, 45
imbalances of, 26, 32, 34
intentions and, 26–28
pronouns and, 44
speed and, 25–26, 38
"thank you" and, 43
Trust and Power Matrix, 25–26, 38
power plays, 22–23, 28, 31, 42
PowerPoint, 61, 124, 183, 262
priority, communication medium choice
as measure of, 48–54
Project Implicit (nonprofit), 166
pronouns, use of, 44
proofreading, 14, 89, 93–94, 110, 168,
230, 236, 244
Ps, 5 (purpose, participants, probable
issues, product, process), 238–239
punctuation
apologies and, 55, 59, 67–69
capitalization and ALL CAPS, 21, 59,
68–70, 113
communication styles and, 56–57
exclamation points, 57–60
gender and, 60, 167–169, 172–175,
176

as measure of emotion, 55–70
periods, 11, 47, 57, 64–65, 175, 188
question mark, 67–68, 213
See also emojis

question mark, 67–68, 213. *See also*
punctuation

racial profiling, xii
remote work
Covid-19 pandemic and, 6, 155,
198–199
digital natives and, 189
digital watercooler moments, 154–155
entirely remove workforce, 118
global remote teams, 210
inclusion and recognition techniques,
16, 95
preventing isolation and
disconnection, 221, 234
René, Danielle, 36
Reply All, 21, 44, 48, 74–77, 110, 137,
138, 230
respect
response time as measure of, 9, 70–74
Value Visibly and, 13–14, 83, 85–89,
92–93, 95–101
response time
acceptable response times, 73
active listening and, 92
anxiety and, 39–41, 71, 74
colleague types and, 244
expectations for, 237, 242
gender and, 168
as measure of respect, 9, 70–74
norms by communication channel,
122–123
patient responses, 137–139
trends in, 229
Roberts, Robin, xiii
Rodney, Denene, 213
Rorschach, Hermann, 30
Rorschach test, 30–31
Rubio, Jen, 147

Sadasivan, Sunil, 182
salutations and sign-offs, 111
common interpretations of, 228–229
culture and, 203, 205, 213–216
formality and, 43–45
gender and, 167, 174

Erica Dhawan is the world's leading authority on twenty-first-century Collaboration and Connectional Intelligence.

Through keynote speaking, training, and consulting, she challenges audiences and organizations to innovate further and faster, together.

She is also the founder & CEO of Cotential, a global organization that helps companies, leaders, and managers leverage twenty-first-century collaboration skills and behaviors to achieve game-changing performance.

She is the co-author of the bestselling book *Get Big Things Done: The Power of Connectional Intelligence*, named #1 on What Corporate America Is Reading. Dhawan was named by Thinkers50 as "The Oprah of Management Ideas" and featured as one of the Top 30 Management Professionals around the world by Global Gurus. She hosts the award-winning podcast *Masters of Leadership*.

Erica speaks on global stages, from the World Economic Forum at Davos and TEDx to companies such as Coca-Cola, FedEx, Goldman Sachs, Walmart, SAP, and Cisco. She writes for *Harvard Business Review*, *Forbes*, and *Fast Company*, and she has degrees from Harvard University, MIT Sloan, and the Wharton School.

Join the community at ericadhawan.com/digitalbody language. Follow the hashtag #digitalbodylanguage.

Follow Erica on LinkedIn: linkedin.com/in/ericadhawan and on Twitter and Instagram @ericadhawan.